I0755853

Narrative Visions of the Willowbrook State School

An Artistic Survey in Bioethics and Special Education

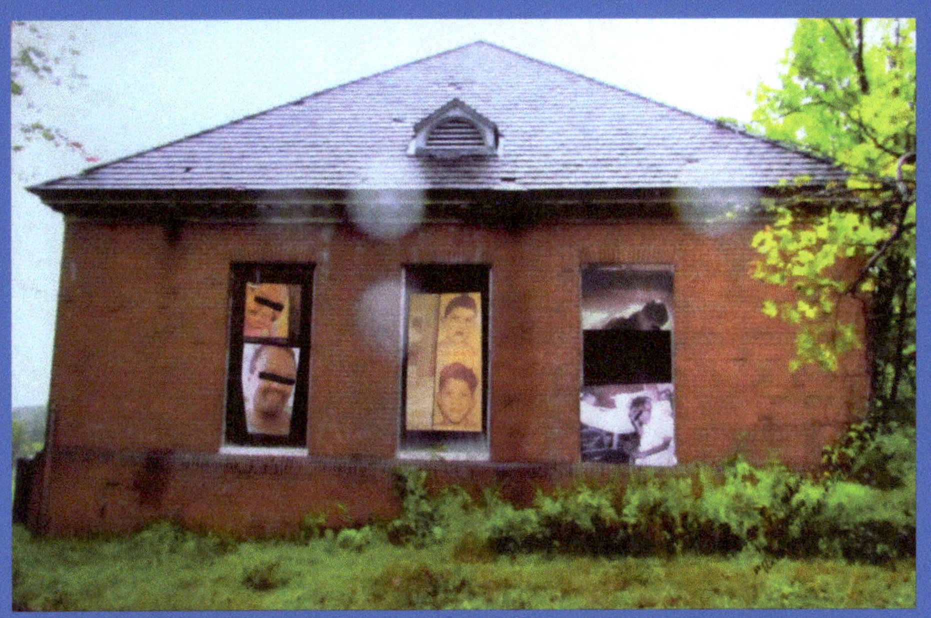

Obiora N. Anekwe,

M.Ed., Ed.D., M.S. Bioethics, M.S.T.

Ethically Speaking Press

Ethically Speaking Press

ISBN: 978-0-578-18167-7

Library of Congress Control Number: 2016947602

PRINTED IN THE UNITED STATES OF AMERICA

Dedicated to the memory of the beloved residents of Willowbrook State School

ABOUT THE BOOK COVER

The artwork entitled, *Unheard Voices of Willowbrook,* is a mixture of a building photograph and visual images I photographed at an exhibition about Willowbrook. The abandoned building in this photograph apparently once housed one of Willowbrook's facilities. Its windows are now boarded up in dark brown wood. The day's dreariness made the impact of taking photographs of the building even more dramatic and resonating. A few days after conducting research at the College of Staten Island, I began to reflect on ways in which I could tell the story of Willowbrook. As I contempated various ideas, I thought that a mixed media photograph depicting the abandoned building would shed light on the disabled men, women, and children who once called Willowbrook home.

In this artistic image, I also highlight how, in many instances, our most vulnerable black and Latino children were unethically treated within the Willowbrook facility. In the collage image, we can visually see this old abandoned building which may have medically treated and housed residents at Willowbrook. At the left side boarded window, I placed photographs of two older African America elders (male and female) who once lived at Willowbrook. I blackened their faces in order to emphasize the concept of confidentiality. Through their hollowed and blackened eyes, I hear the tales of a thousand ancestral voices. But within the middle boarded window, I decided to leave two photographs of a young Latino boy's eyes visible in order to shed light on the lack of humanity shown toward young people. The right hand boarded window in the image shows how young children were systematically impacted by the unethical medical studies at Willowbrook. I left the middle area of this right-sided boarded window empty in order to symbolize the voided space of victims' unheard voices affected by Willowbrook.

Now, as I think of Willowbrook and all the countless mistakes that were made there, I am reminded that we, as educators and bioethicists, are called to serve those in greatest need, to right the wrongs of the past, and to go forth with the greatest promise of our calling!

Obiora N. Anekwe
c. 2014
PHOTOGRAPH/MIXED MEDIA ON PAPER
16 by 24 INCHES

Original image of building photographed by the artist.

Additional images provided by the College of Staten Island's Archives and Special Collections.

Table of Contents

FOREWORD

Bioethics and special education have inextricably been connected to one another in the historical context and study of justice and universal civil rights in the American cultural landscape. These two areas have more in common than most people would have considered, symmetrically rooted in the egregious history affiliated with unethical and non-consensual human research experimentations conducted on children with intellectual disabilities. Historically, the Willowbrook State School (1947-1987), a state-supported institution for children with intellectual disabilities, once located in the Willowbrook neighborhood of Staten Island, New York, represents the best example of the intersection between bioethics and special education studies. The environmental and living conditions for residents with special needs at Willowbrook were so unhealthy and unsanitary that Senator Robert Kennedy described the facility as a snake pit after he toured the institution in 1965.

Further evidence of these deplorable conditions were exposed several years later through an exposé in early 1972 when Geraldo Rivera, then an investigative reporter for WABC-TV in New York, conducted a series of investigations at Willowbrook, uncovering a host of deplorable conditions, including overcrowding, inadequate sanitary facilities, and physical and sexual abuse of residents by members of the school's staff. The exposé, entitled *Willowbrook: The Last Great Disgrace*, received national attention, even winning a Peabody Award.

Described by many bioethicists as one of the most unethical medical experiments ever performed on children in the United States, Willowbrook is not only studied in special education studies, but in the field of bioethics due to its impact on framing policies and laws protecting minors and other vulnerable populations in human subject experimentations. Throughout the first decade of the institution's operation, outbreaks of hepatitis A were common at the school due primarily to unsanitary living conditions. Eventually, these outbreaks laid the foundation for medical researchers to conduct controversial medical studies on residents with special needs at Willowbrook between the late 1950s and the 1970s. These researchers, primarily Saul Krugman (New York University) and Robert W. McCollum (Yale University), monitored subjects to gauge the effects of gamma globulin in combating hepatitis A. A public outcry forced the research project and medical studies to be discontinued after accusations were leveled that these researchers had used mentally disabled children as "human guinea pigs" in unethical clinical trials without proper informed consent.

In addition to the intersection between bioethics and special education, the evaluation of these two areas of study can be measured through art. My goal for this book is to uniquely express and chronicle the stories of Willowbrook State School through expressive art collages and photographs created from my own perspective as a bioethicist and special education teacher. Within most of the collages and mixed media pieces throughout my book, I depend heavily on using various forms of geometry and geometric renderings that are reminiscent of artist, Romare Bearden, who is one of my greatest influences as an artist. Since 2014 and the summer of 2015, I have created more than forty art collages, mixed media images, and photographs that vicariously express the impact of Willowbrook to the larger world. My book consists of two main chapters: photographs and collages/mixed media images about the Willowbrook State School. This book has served as an opportunity for me to help the larger global community understand the impact of injustice and unethical human subject experimentations conducted on populations with special needs. In some transformative way, I also hope that this book tells the story of countless children with special needs who were negatively affected by Willowbrook. Hopefully, my artwork and photographs will begin to repair the wrongdoings and restore faith back to a medical community who desires to positively impact scientific discovery through humane and balanced research experimentation.

1

The Willowbrook Photography Series

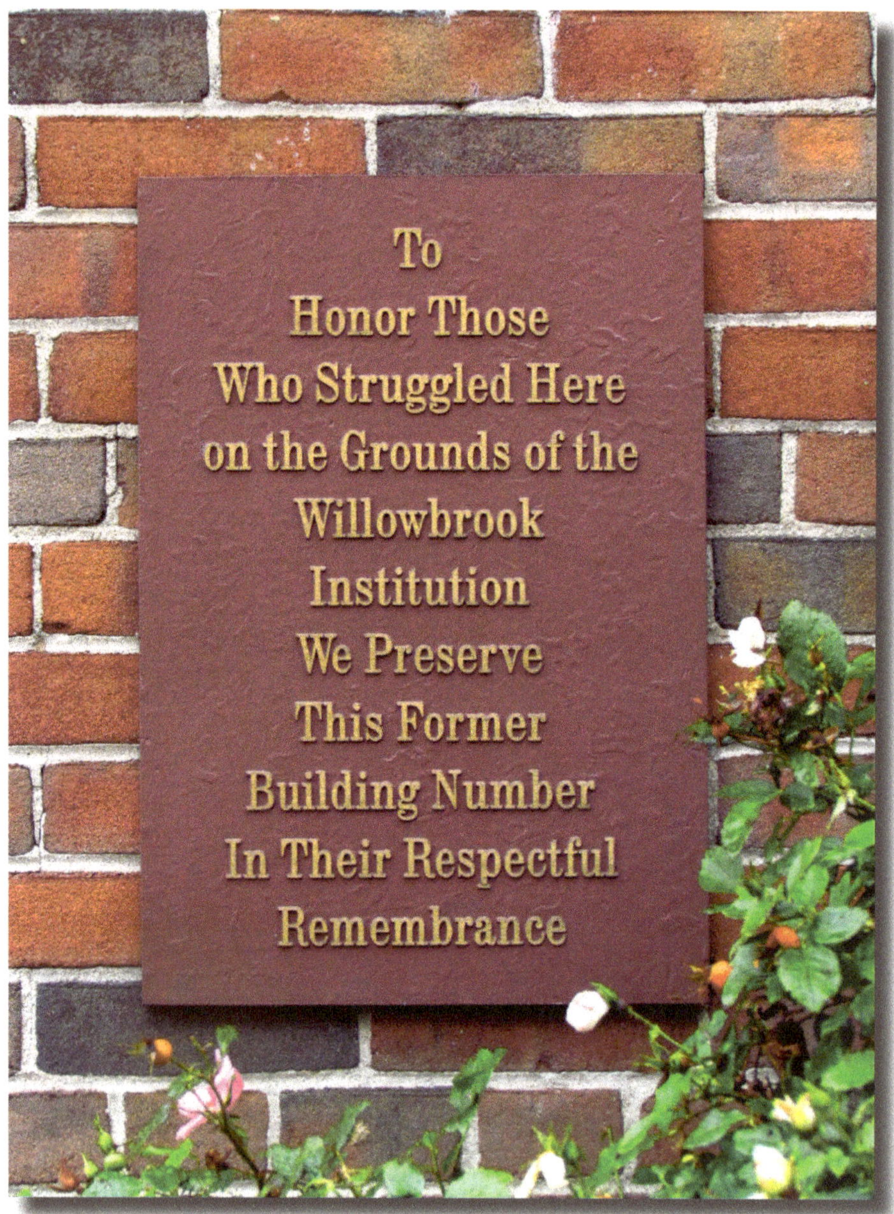

Historic marker commemorating Willowbrook State School, Staten Island, New York

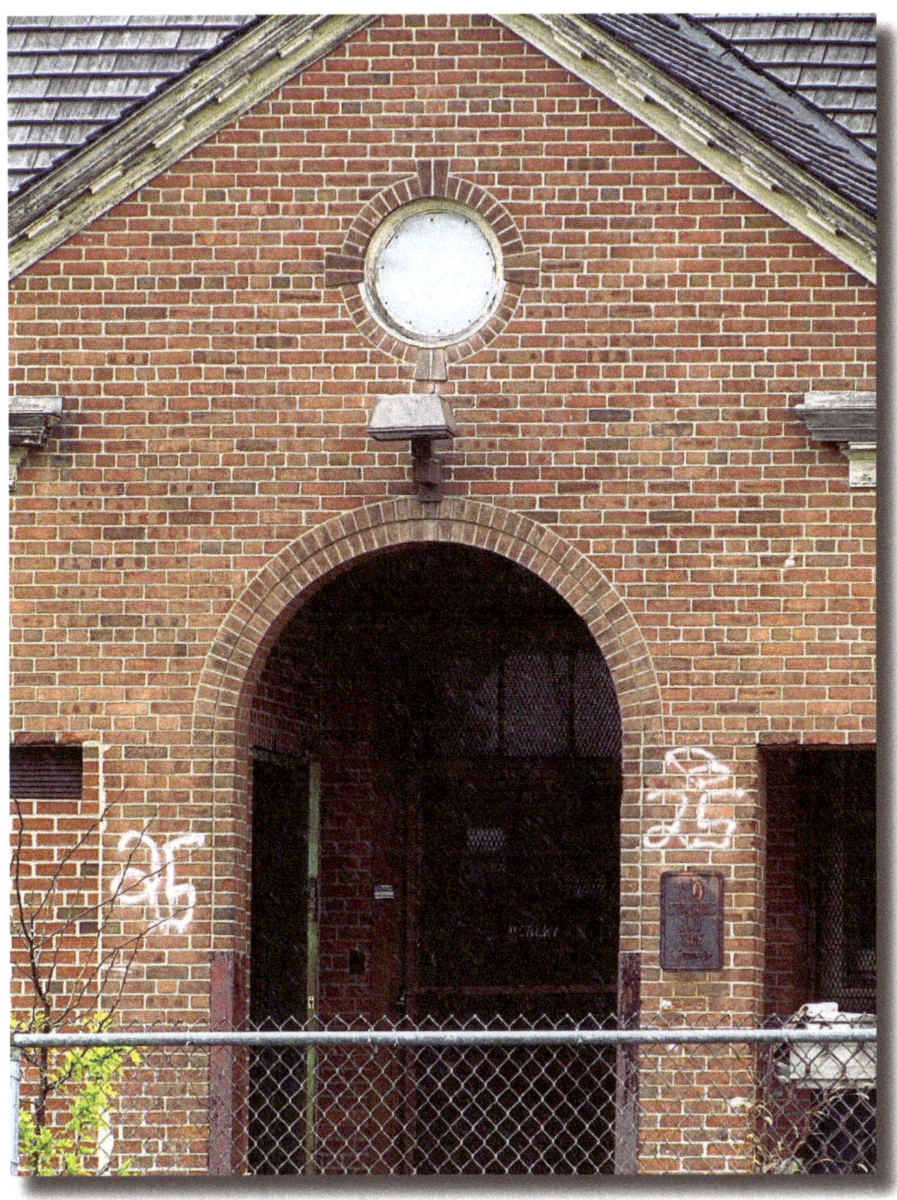

Gated abandoned building at the former Willowbrook State School, Staten Island, New York

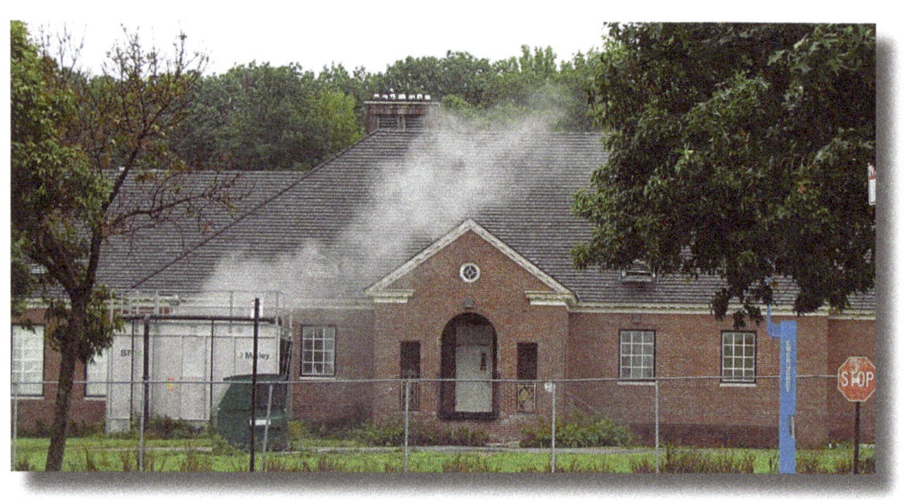

Wide image of gated abandoned building at the former Willowbrook State School, Staten Island, New York

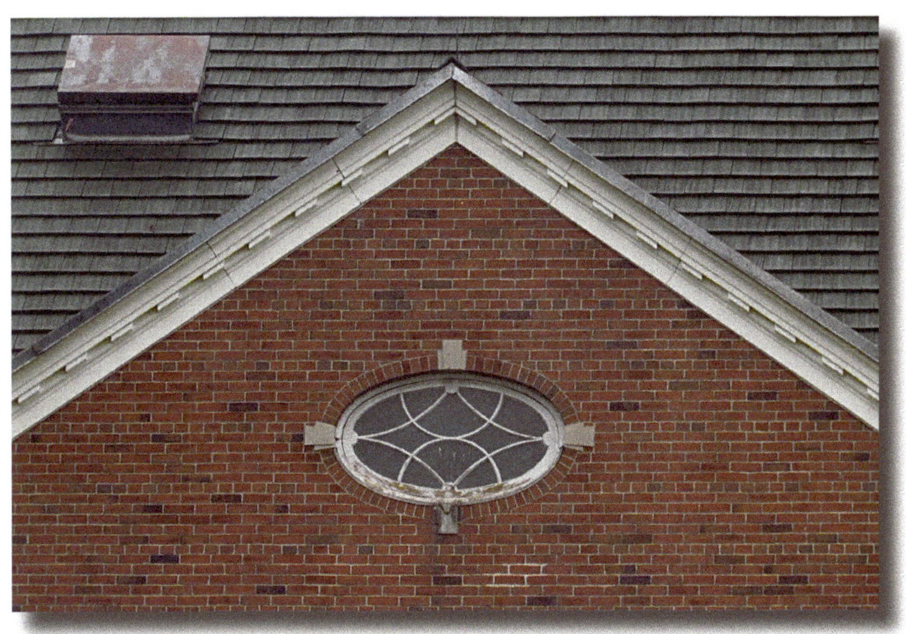

Front image depicting top triangle view of building's window at the former Willowbrook State School, Staten Island, Now York

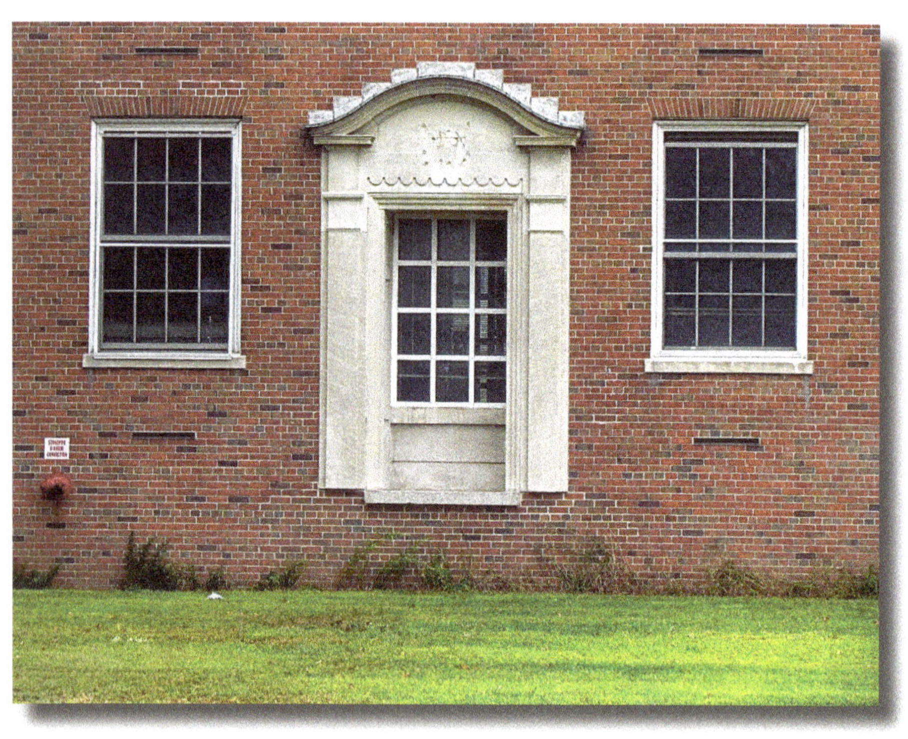

Front image of abandoned building with twin windows and door at the former Willowbrook State School, Staten Island, New York

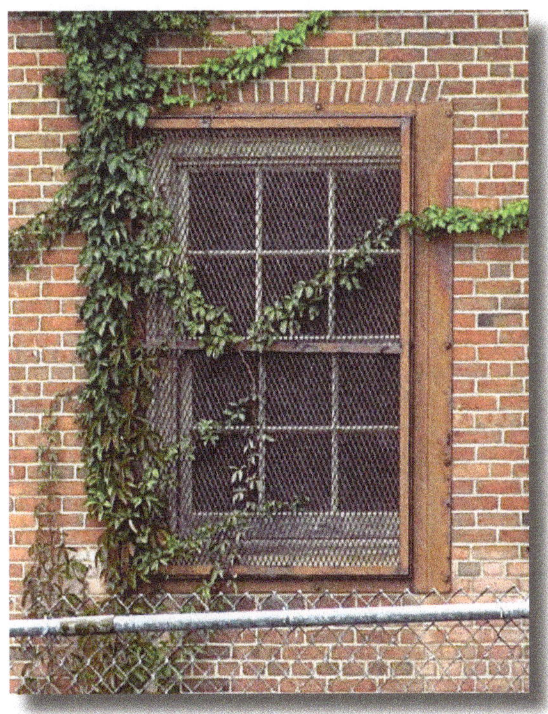

Close-up view of secured window of abandoned building at the former Willowbrook State School, Staten Island, New York

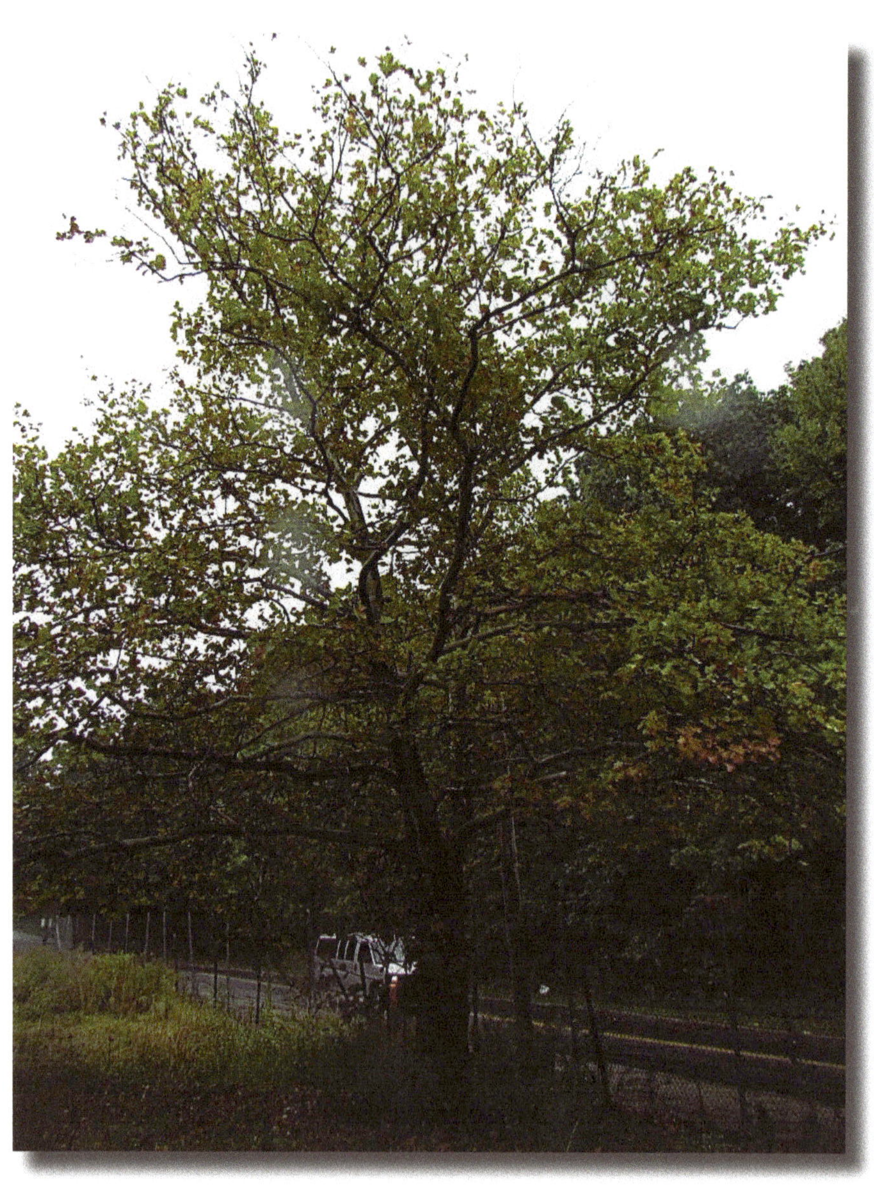

Large tree near abandoned building at the former Willowbrook State School, Staten Island, New York

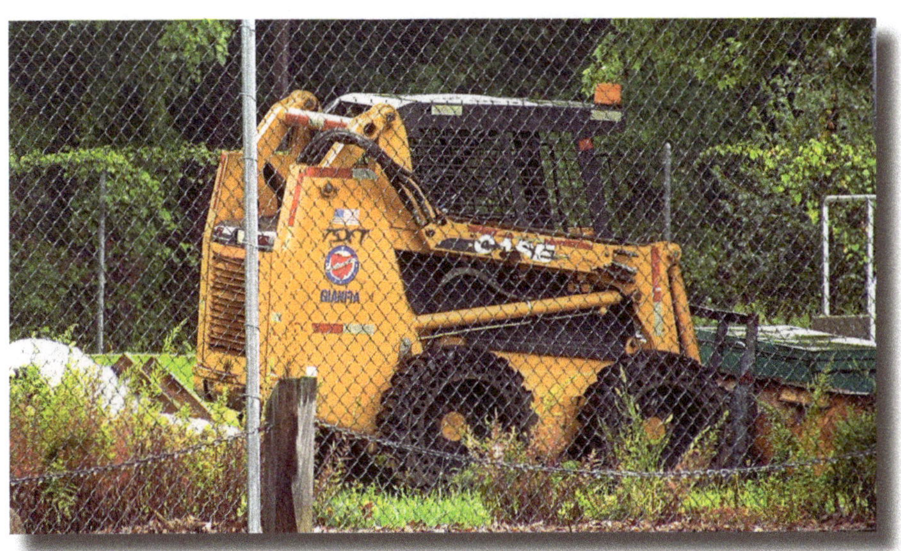

Large movable tracker near abandoned building at the former Willowbrook State School, Staten Island, New York

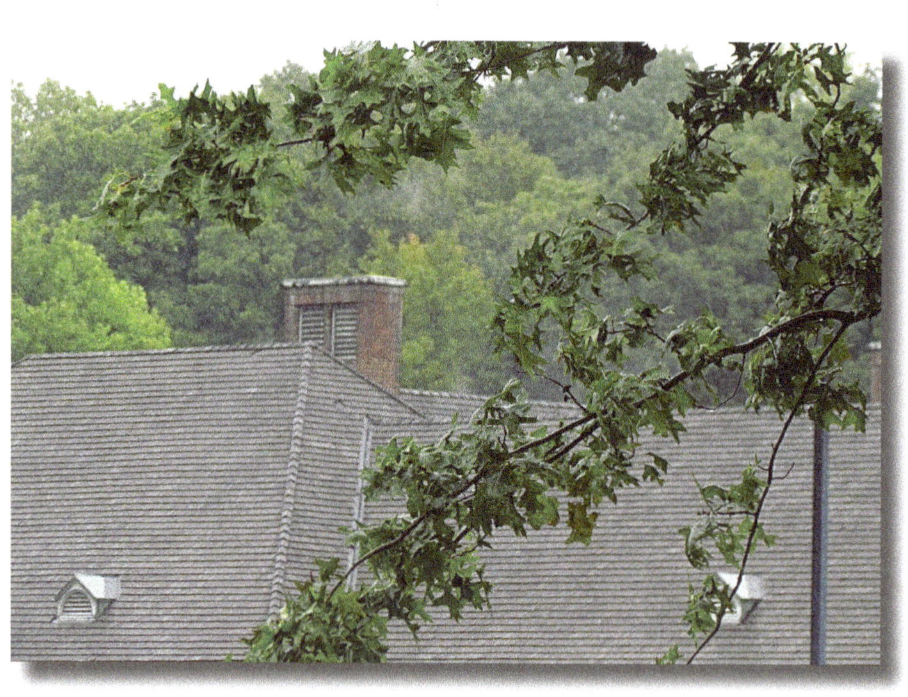

Image view of building's chimney at the former Willowbrook State School, Staten Island, New York

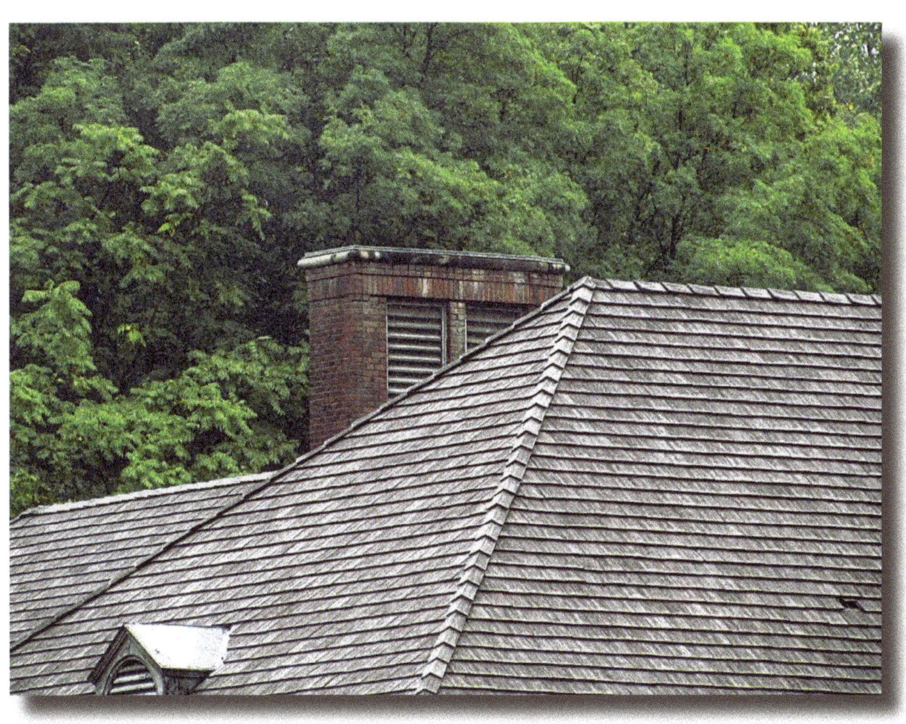

Close-up view of building's chimney at the former Willowbrook State School, Staten Island, New York

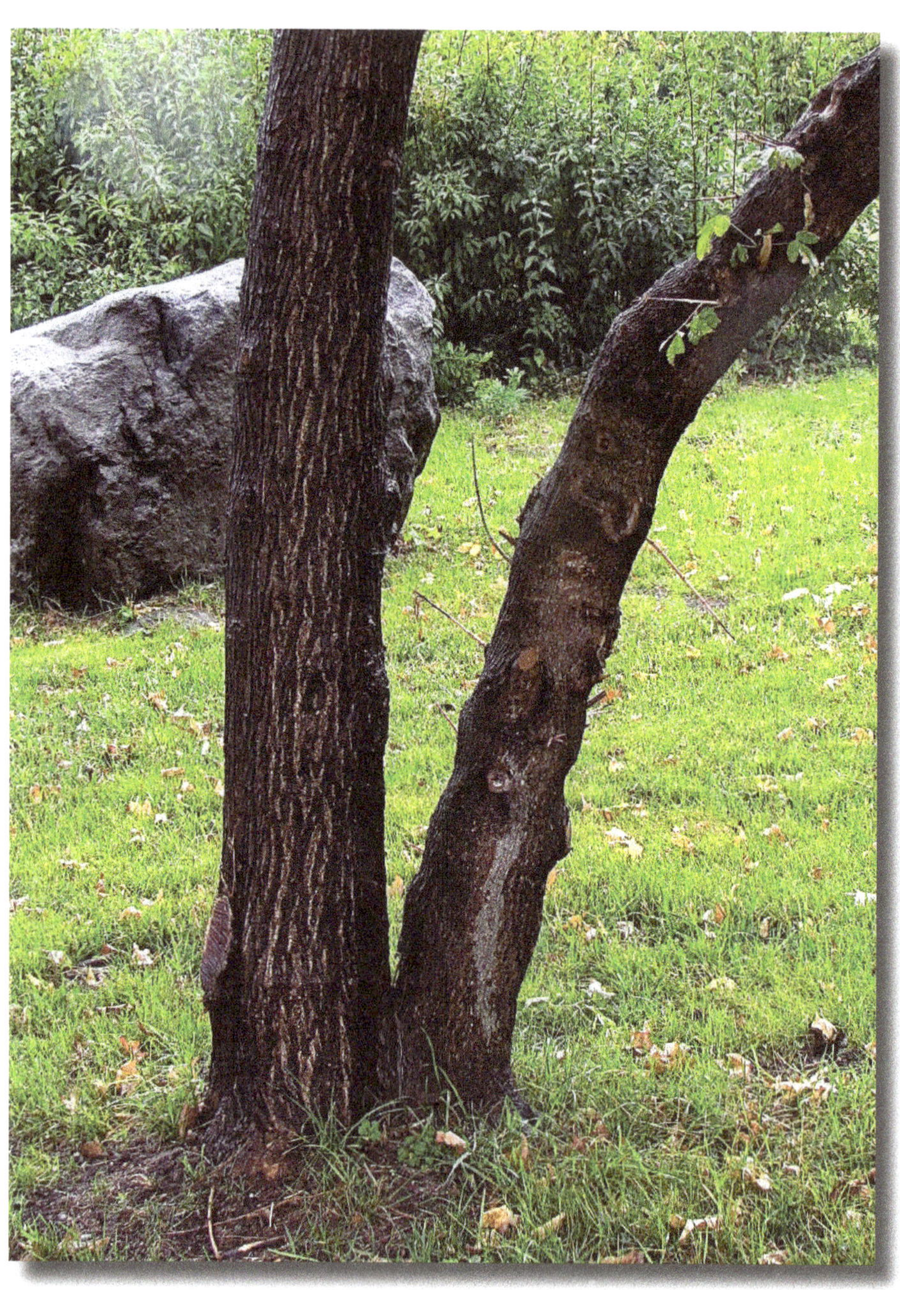

Image of tree located at the former Willowbrook State School, Staten Island, New York

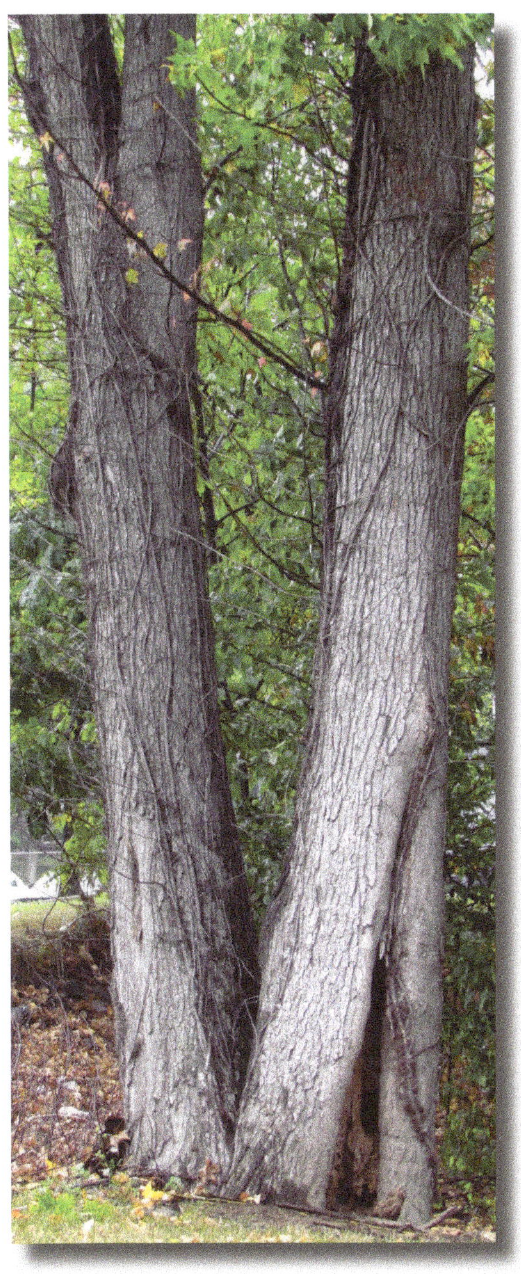

Image of tree located at the former Willowbrook State School, Staten Island, New York

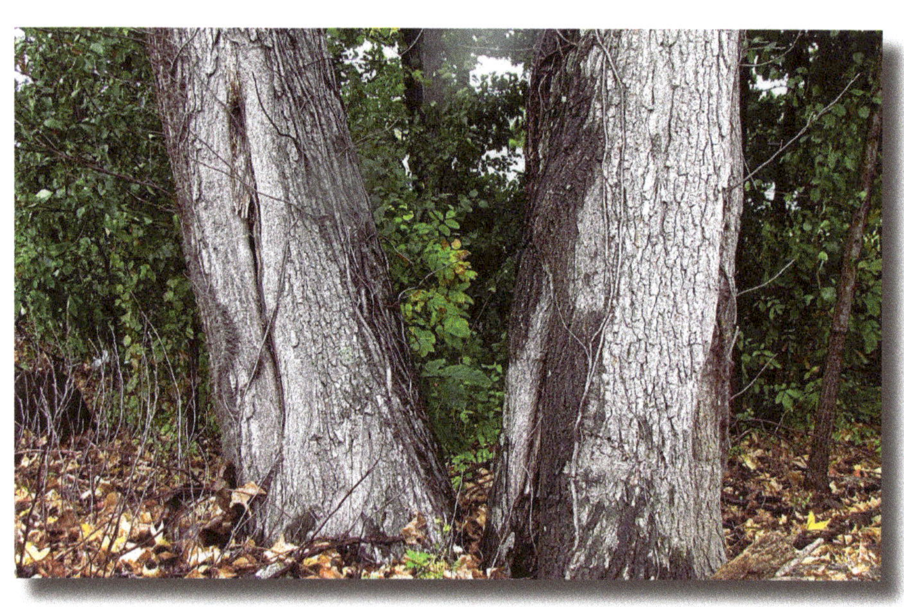

Close-up image of tree located at the former Willowbrook State School, Staten Island, New York

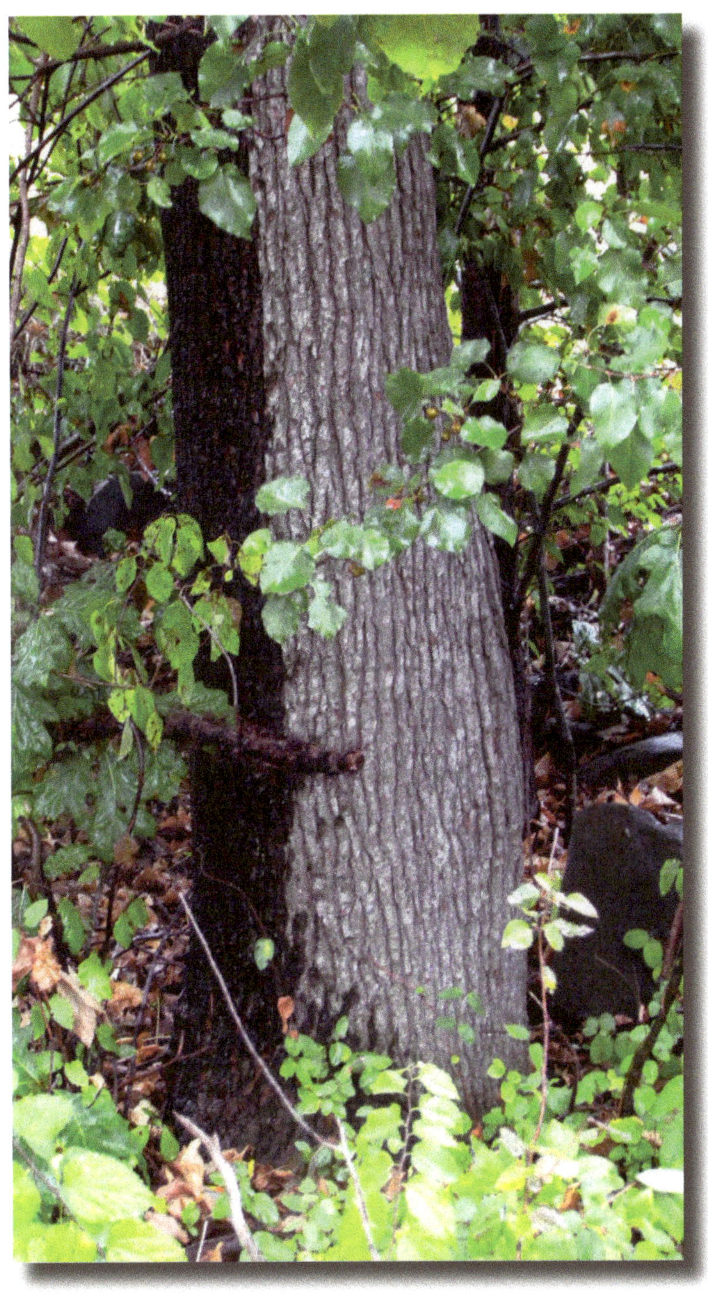

Close-up image of tree located at the former Willowbrook State School, Staten Island, New York

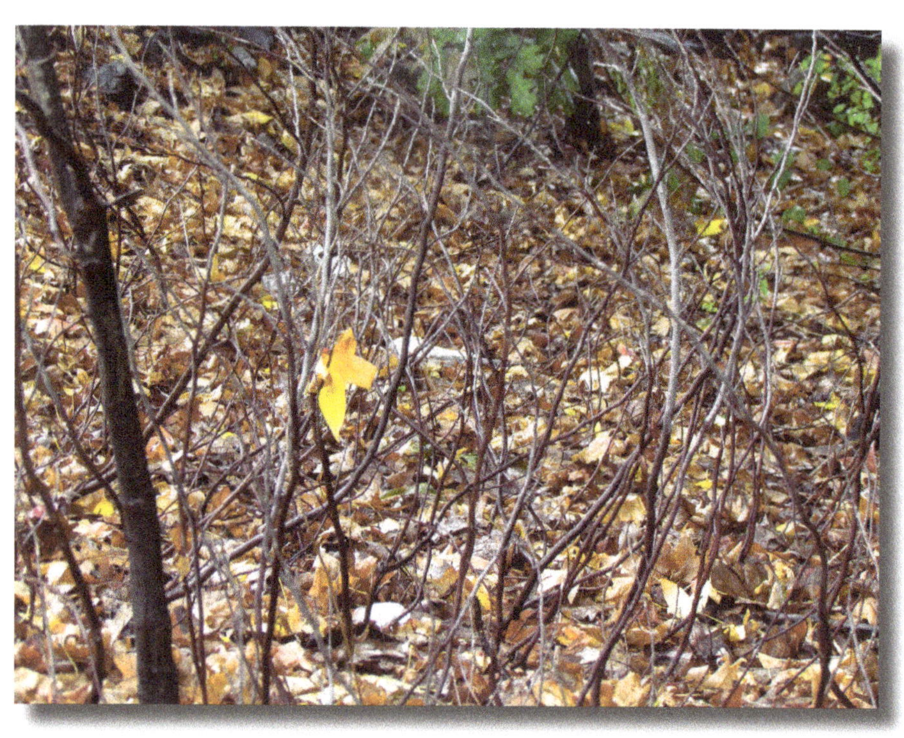

Close-up image of leaves at the former Willowbrook State School, Staten Island, New York

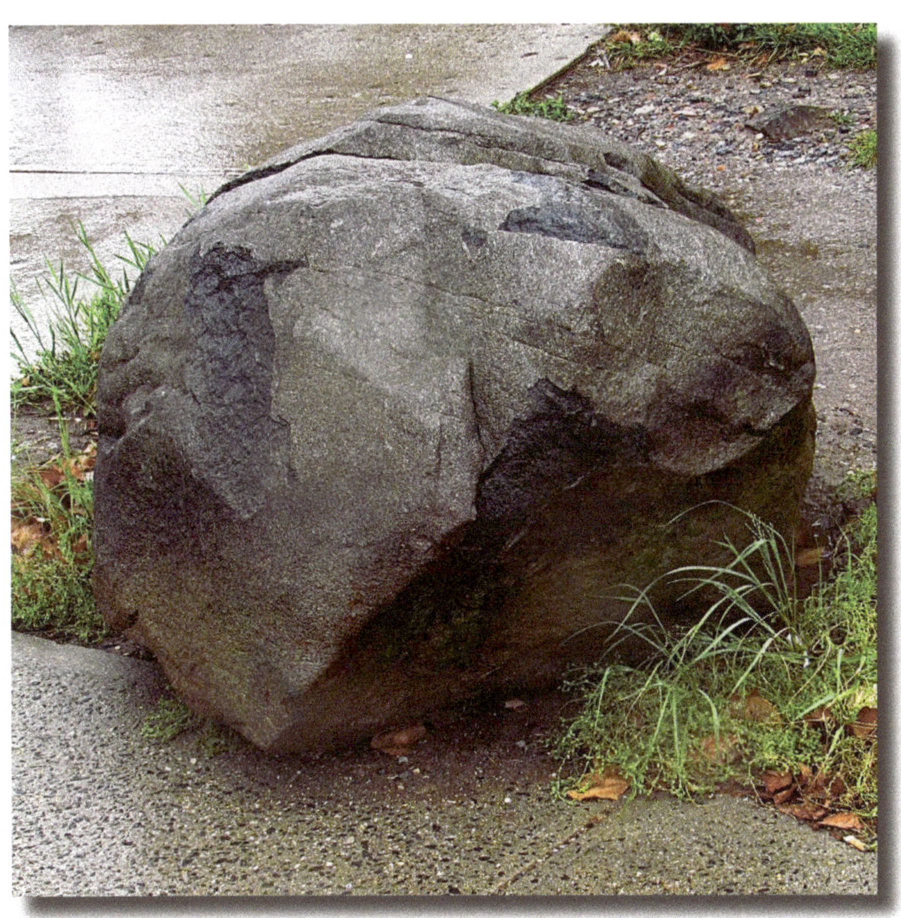

Image of aged stone at the former Willowbrook State School, Staten Island, New York

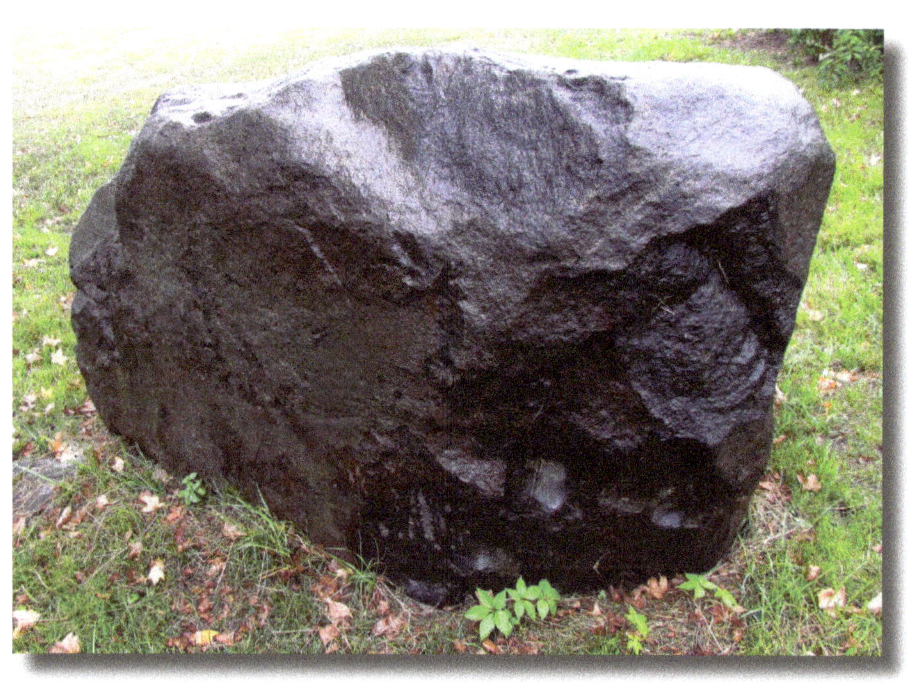

Image of aged stone at the former Willowbrook State School, Staten Island, New York

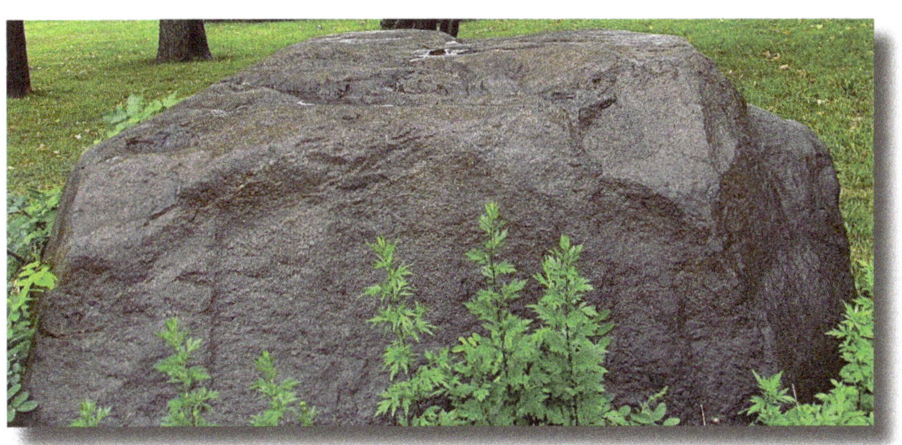

Image of aged stone at the former Willowbrook State School, Staten Island, New York

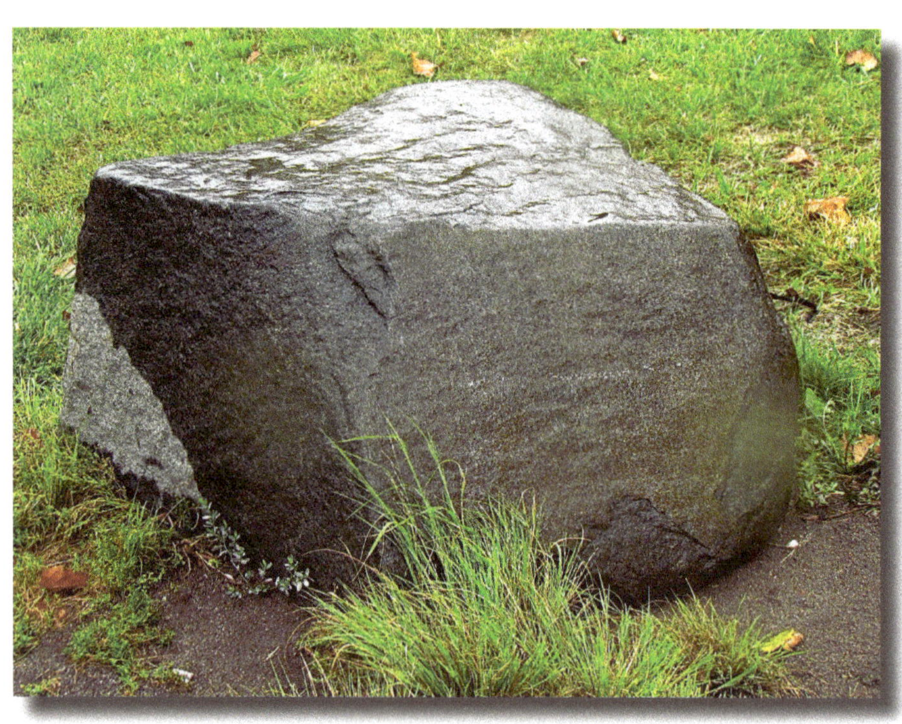

Image of aged stone at the former Willowbrook State School, Staten Island, New York

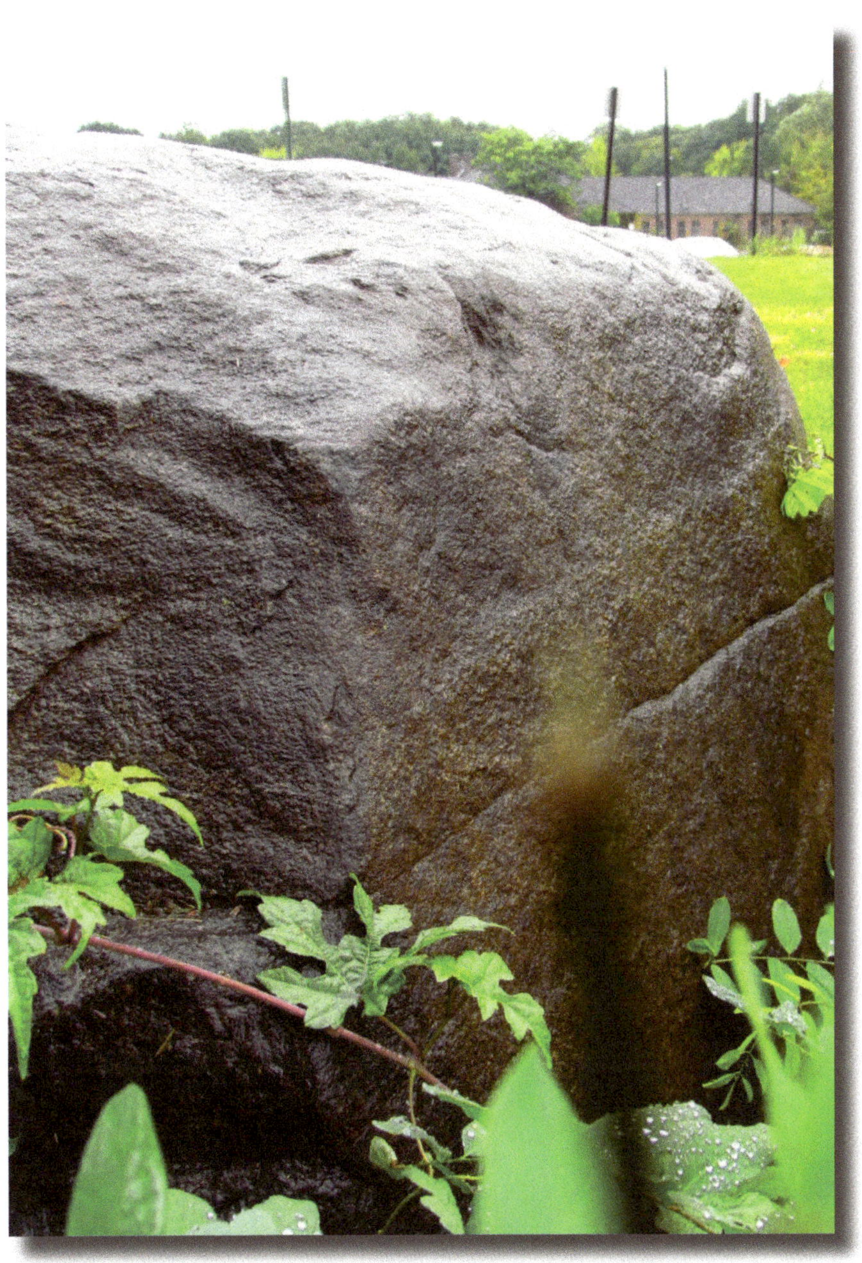

Image of aged stone at the former Willowbrook State School, Staten Island, New York

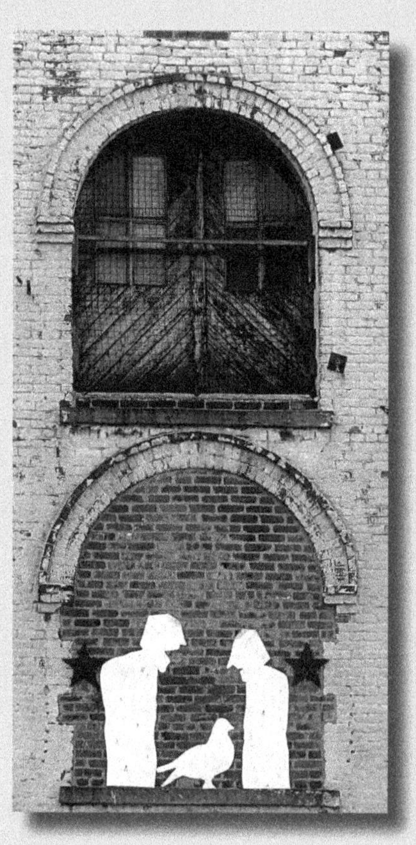

Ancestral Messenger II

Since residing in Brooklyn, New York, I have noticed visible connections between ethics and art. One day while walking in the Crown Heights neighborhood back home to Bedford-Stuyvesant in Brooklyn, New York, I noticed an old warehouse-like building on the corner of Franklin Avenue and Bergen Street. In the enclosed brick window on the second floor, I saw an artistic scenic image seemingly depicting two elderly men talking to one another, separated only by a stationary bird. The image somehow reminded me of an art rendering familiar to southern folk art. As I observed the figurative work, I thought about the students at Willowbrook who could not speak for themselves because of their mental and/or physical disabilities. I wondered to myself what words or sounds these individuals communicated to each other while they resided at Willowbrook as patients and residential students.

Entitled, *Ancestral Messenger II,* I decided to photograph the image I saw on the building in Brooklyn because I did not want the memory of this scene to escape my mind. I wanted it to serve as a permanent reminder of how art interprets the stories of a people. As I further analyzed the image, I thought about how, in many ways, the bird at the center of the men's conversation served as a constant reminder that there will always be a spiritually guided voice from a messenger who is destined to carry the stories, conversations, and experiences of a people to the next generation so that we may know who we are and who we are destined to become.

Obiora N. Anekwe
c. 2015
PHOTOGRAPH

2

Collages and Mixed Media on Paper

HISTORIC WILLOWBROOK STATE SCHOOL BUILDINGS

The buildings at Willowbrook State School represent the physical structure and symbolic nuance of the institution. During my initial research studies about Willowbrook, I visited the College of Staten Island, which is located on the former campus of Willowbrook State School. I conducted research in the archives office at the college and received several file images of Willowbrook's buildings from the archivist. After reviewing these photographs several days later, I knew that I could eventually use these images in developing several collages highlighting the physical structure of the Willowbrook campus.

My five building collages are earth tone in color and mixed with found paper/mixed media paper. One particular building collage entitled, *Willowbrook Building #5*, highlights a photograph I took of the moon. This collage is of particular interest to me because it fits so well into the color scheme and aura of my vision for the piece. All in all, I enjoyed creating these five small collages because they actually show what some of the buildings on the Willowbrook campus actually looked like, but with an expressionistic twist.

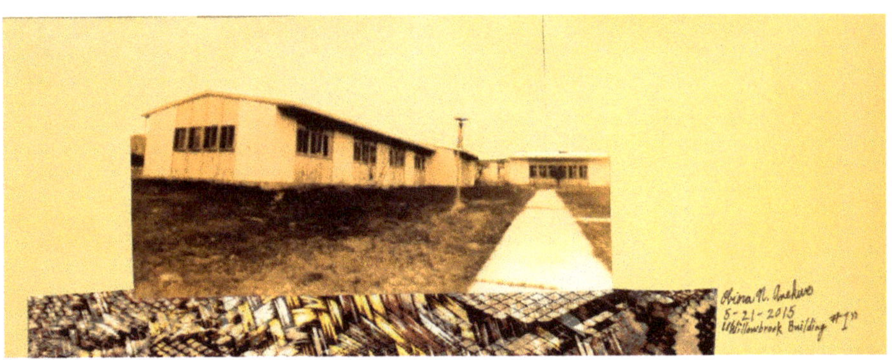

WILLOWBROOK BUILDING #1

c. 2015. COLLAGE ON PAPER, 7 by 21 INCHES

Original image of building provided by the College of Staten Island's Archives and Special Collections

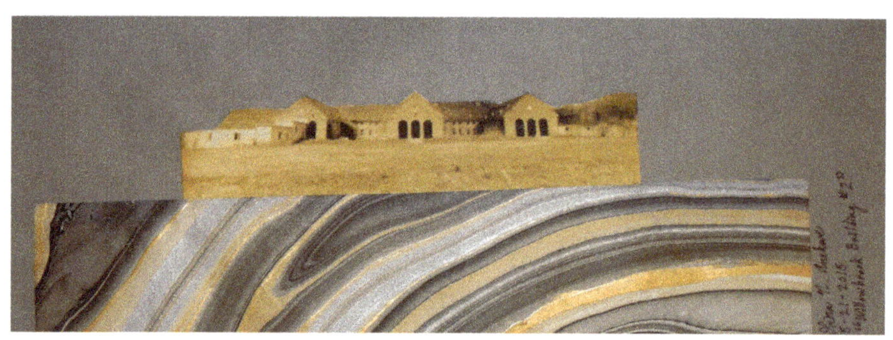

WILLOWBROOK BUILDING #2
c. 2015. COLLAGE ON PAPER, 7 by 21 INCHES

Original image of building provided by the College of Staten Island's Archives and Special Collections

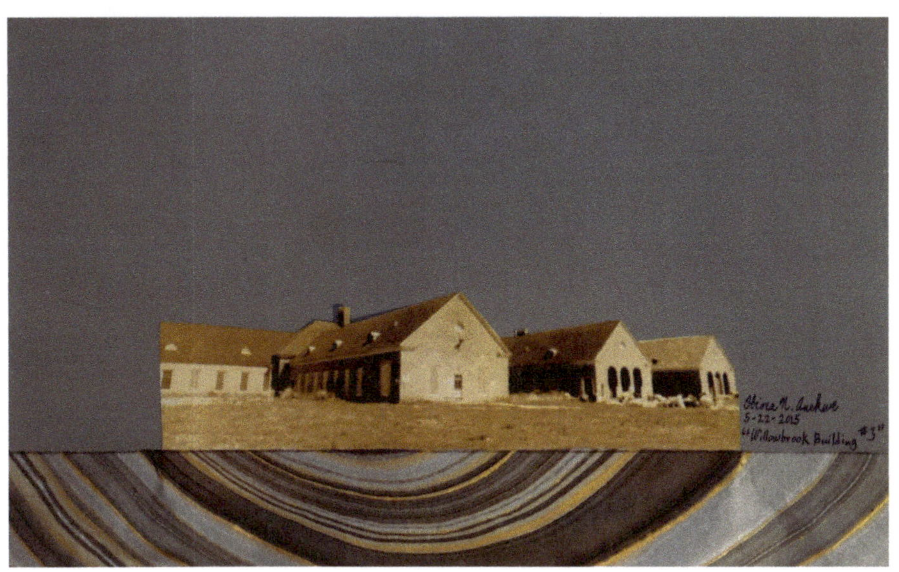

WILLOWBROOK BUILDING #3

c. 2015. COLLAGE ON PAPER, 11 by 15 INCHES

Original image of building provided by the College of Staten Island's Archives and Special Collections

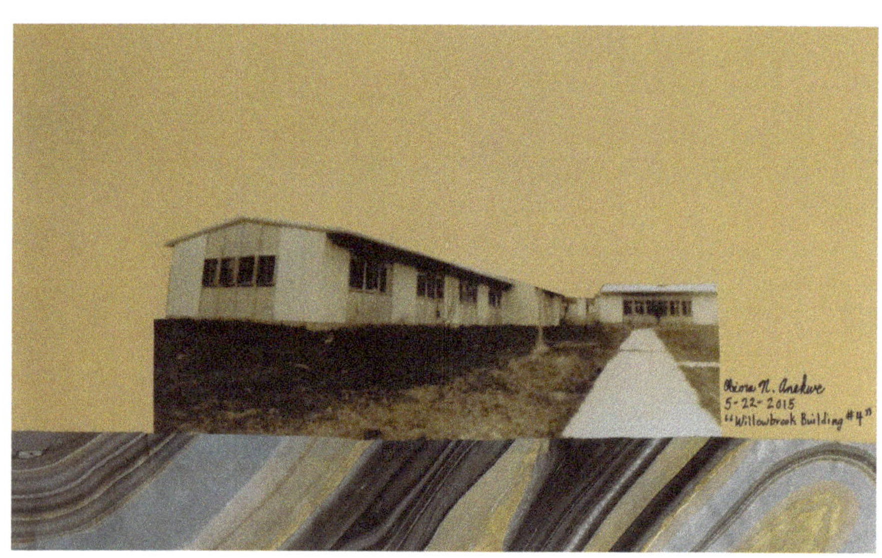

WILLOWBROOK BUILDING #4

c. 2015. COLLAGE ON PAPER, 11 by 15 INCHES

Original image of building provided by the College of Staten Island's Archives and Special Collections

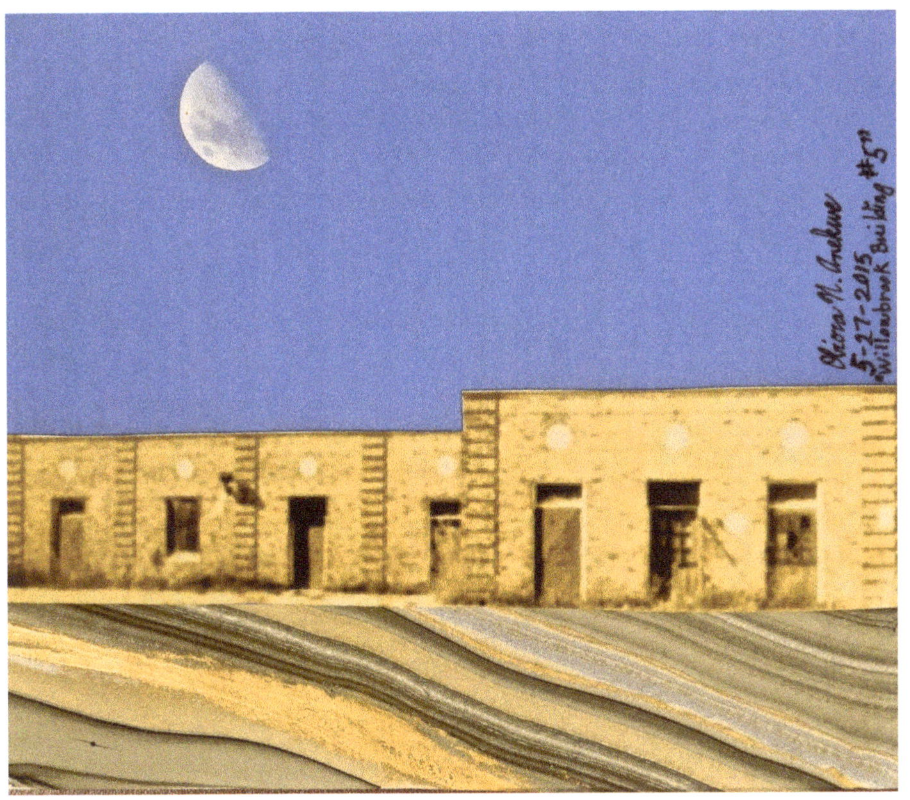

WILLOWBROOK BUILDING #5

c. 2015. COLLAGE ON PAPER, 7 by $8^{1/2}$

Original image of building provided by the College of Staten Island's Archives and Special Collections

THE UNIVERSAL SPIRITUALISM OF WILLOWBROOK

The following art collages represent the spiritual aspects associated with the tragic and unethical events that occurred at the Willowbrook State School. Some of the collages included in this chapter: *Hands of God, The Dream Lives Within Me, All Rise!, Angels of Willowbrook,* and *Let Us Pray* are all works of art that I feel directly speak to the spiritual aspects of Willowbrook, which is rarely discussed and analyzed. The title of each artwork really speaks directly to the visual representation of each individual piece. For instance, the collage, *Hands of God,* highlights the fact that we are all responsible for change within the world we live in. We are the change we need in the world and the God we serve moves when we join Him in the pursuit of justice and peace.

In another piece, *Angels of Willowbrook,* I highlight the fact that many children of color were affected and involved in unethical experiments at Willowbrook, especially African American children. In my opinion, these children, especially represented by the three girls in the collage, are angels who serve as martyrs to the cause of informed consent. The background colors of red, yellow, and black in this collage represent the German flag. I wanted to make a comparison in this collage to the European Holocaust of Jewish children, which was similar to how some children were experimented upon at Willowbrook.

The collage, *All Rise!,* speaks to the fact that we, as citizens, are responsible for rising to the occasion of fighting for human justice. We are to essentially rise to the highest occasion in order to fulfill our greatest selves. *The Dream Lives Within Me* reminds us all that every dream and aspiration we have as a people lives within each individual. We are equipped with the skill and ability to accomplish all our dreams if we remain focused and committed to our goals. The surrounding hands that are in the collage come from the hands of prominent dreamers from around the world who made a change in their global community, including Martin Luther King, Jr., Michael Jackson, Malcolm X, and Jorge Cardinal Bergoglio.

In *Chains,* we are reminded that all human beings at some point have felt limited and encaged physically, mentally, or spiritually. Such feelings of despair are not simply limited to what stereotypes society imposes on the disabled, but on everyday citizens who believe they are free, but are unconsciously chained down one way or the other. The collage, *Higher Knowledge*, reminds the reader that although Willowbrook was established as a residential institution of higher learning and care for the disabled, it essentially strayed from its mission and imposed suffering and a lack of sufficient care upon its residents. In the piece, *Reaching Out*, I dramatize the image of the interior surroundings of pain many residents endured at Willowbrook. *Human Suffering* and *Let Us Pray* are two collages that co-exist with each other because they compare and contrast the aspect of universal suffering found in the world. The human figures are representative of men who were refugees and victims of starvation in Eastern Europe during civil war in the 1980s-1990s. *Let Us Pray,* in particular, represents how prayer, in and of itself, changes things, especially when it relates to changing how justice is rendered.

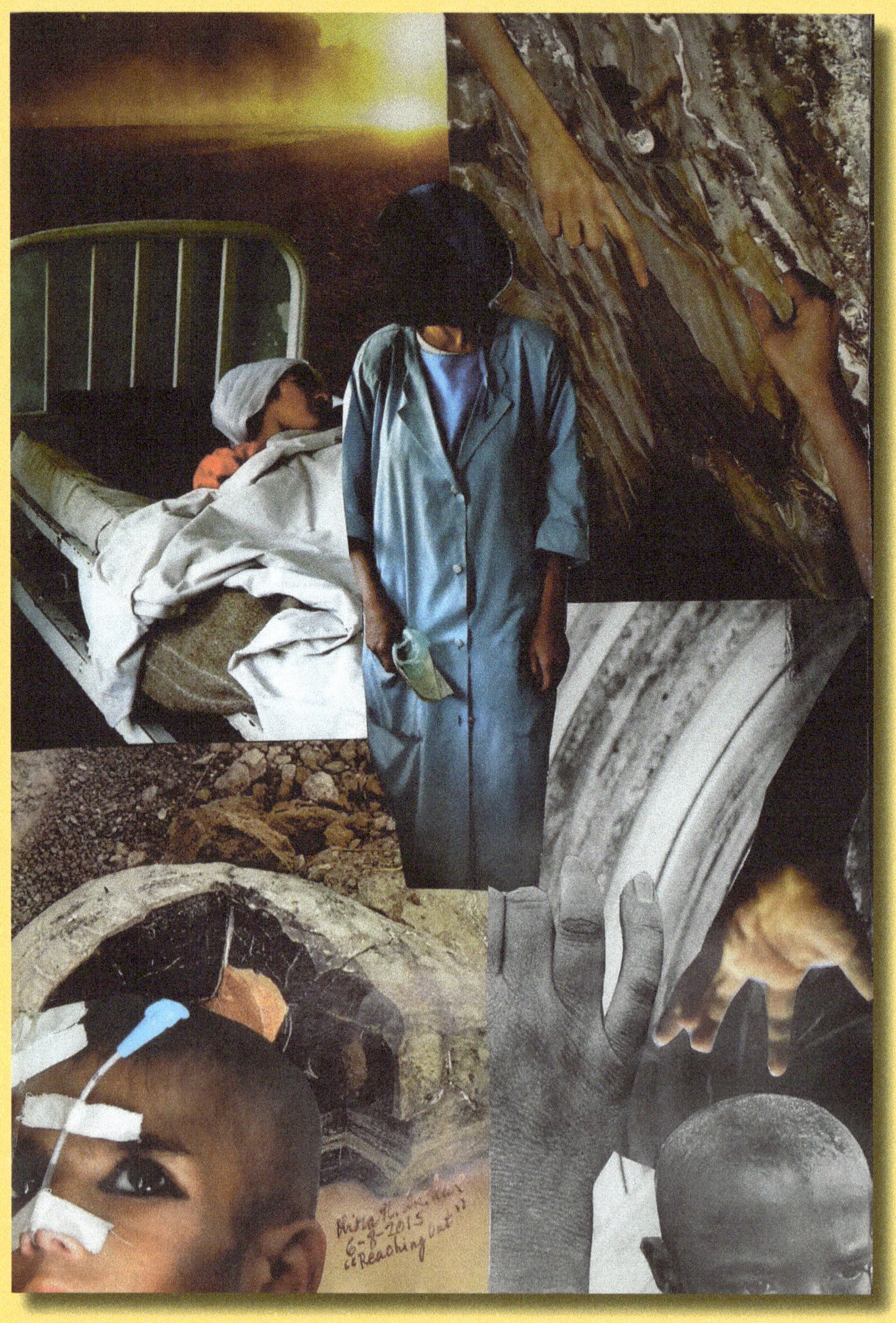
6-8-2015
"Reaching Out"

Reaching Out

The mixed media image, *Reaching Out*, is directly related to residential facilities that house medically-ill patients unethically. One such example is the Willowbrook State School, which housed residents with special needs. With highly unsanitary living conditions, hepatitis A was spread among residents with little to no preventive barriers to aid victims. But more tragically, medical researchers took advantage of residents with hepatitis A and experimented on them without the appropriated merits of informed consent by parents or mentally competent residents. The four exterior areas of the image highlight deprived conditions of similar settings like Willowbrook. The top left side image depicts a sick teenager in a hospital bed with a sunset in the background. The top right hand image shows the hands of two beings visibly attempting to connect, which symbolizes the human connection among people around the world. Thus, we are all humanly connected through the universal suffering we all encounter. The bottom left image shows a young child with medical tubes connected to the frontal head area. This image symbolizes how children often suffer physically, spiritually, and psychologically. Its background image shows the barren earth which also emphasizes how environmental factors affect children's human suffering. In comparison to the top right image, the bottom right image shows two extended hands reaching out for a small boy's head. This image relates to how adults should strive to protect all children, no matter their physical and/or mental disability. In the final center image, the blackened out face of the nurse emphasizes the feelings of despair health care practitioners feel when they are faced with the grave sufferings of children housed in residential facilities.

Obiora N. Anekwe
c. 2015
MIXED MEDIA ON PAPER
11 by 16 INCHES

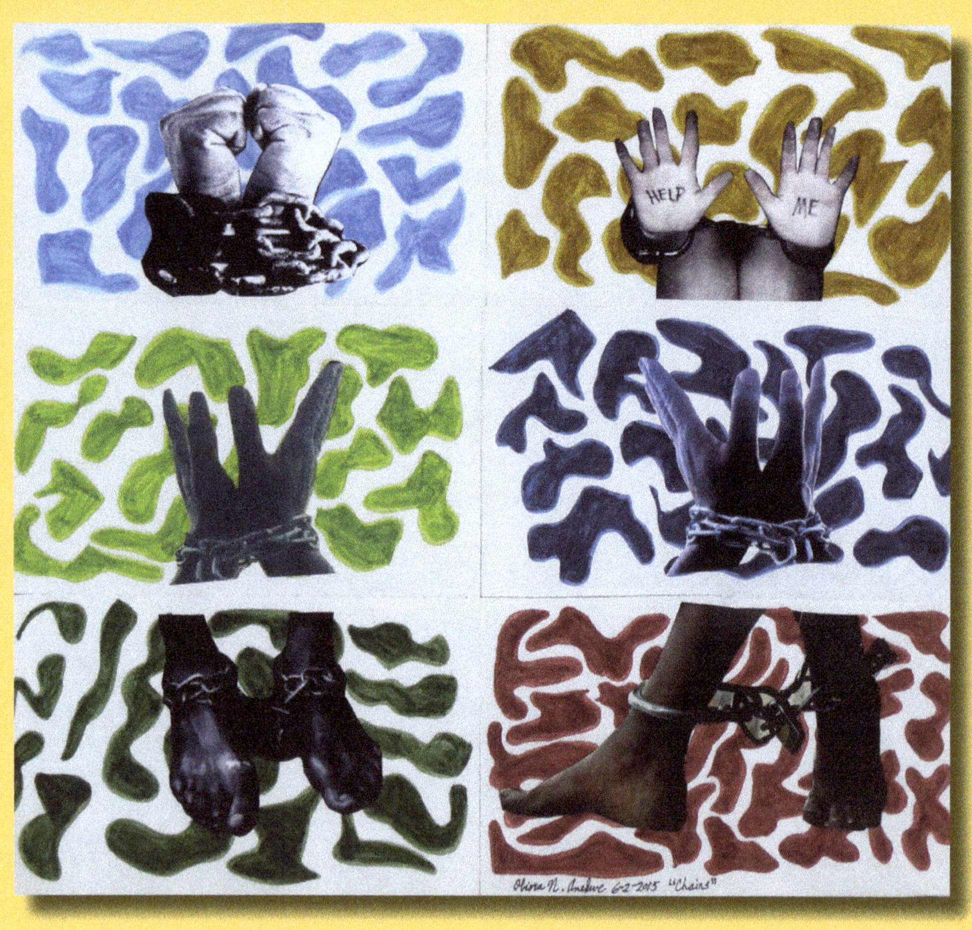
HELP
ME
Obiora N. Anekwe 6-2-2015 "Chains"

Chains

In life, there are literally chains that can keep us locked into challenges that seem to overcome us. In a way, we see no way out. Life's chains can come in the form of several layers, through mental, physical, or spiritual barriers either self-imposed or placed upon us by society. In either case, it is ultimately the responsibility of human beings to lift the chains off one another and bring hope to the most vulnerable. The art piece, *Chains*, represents a re-awakening of the human spirit to visually recognize and actively participate in lifting the veil of discontent among those who suffer the most, which includes young people who have mental and physical disabilities. When there is a cry for help, as depicted in the top right portion of the artwork, we, as a global community, can take the bold move to eradicate injustice wherever it exists. The colored objects in the piece's background represent the unborn who are the future of our human race. Their figurative shapes symbolize the diversity in color, race, shape, size, and depth of possibilities for the unborn human being. Before they are brought into the world, we, as caretakers, serve as light bearers who shine light into their humanity so that they will not have to live under undue suppression and injustice.

Obiora N. Anekwe
c. 2015
MIXED MEDIA ON PAPER
15 by 17 INCHES

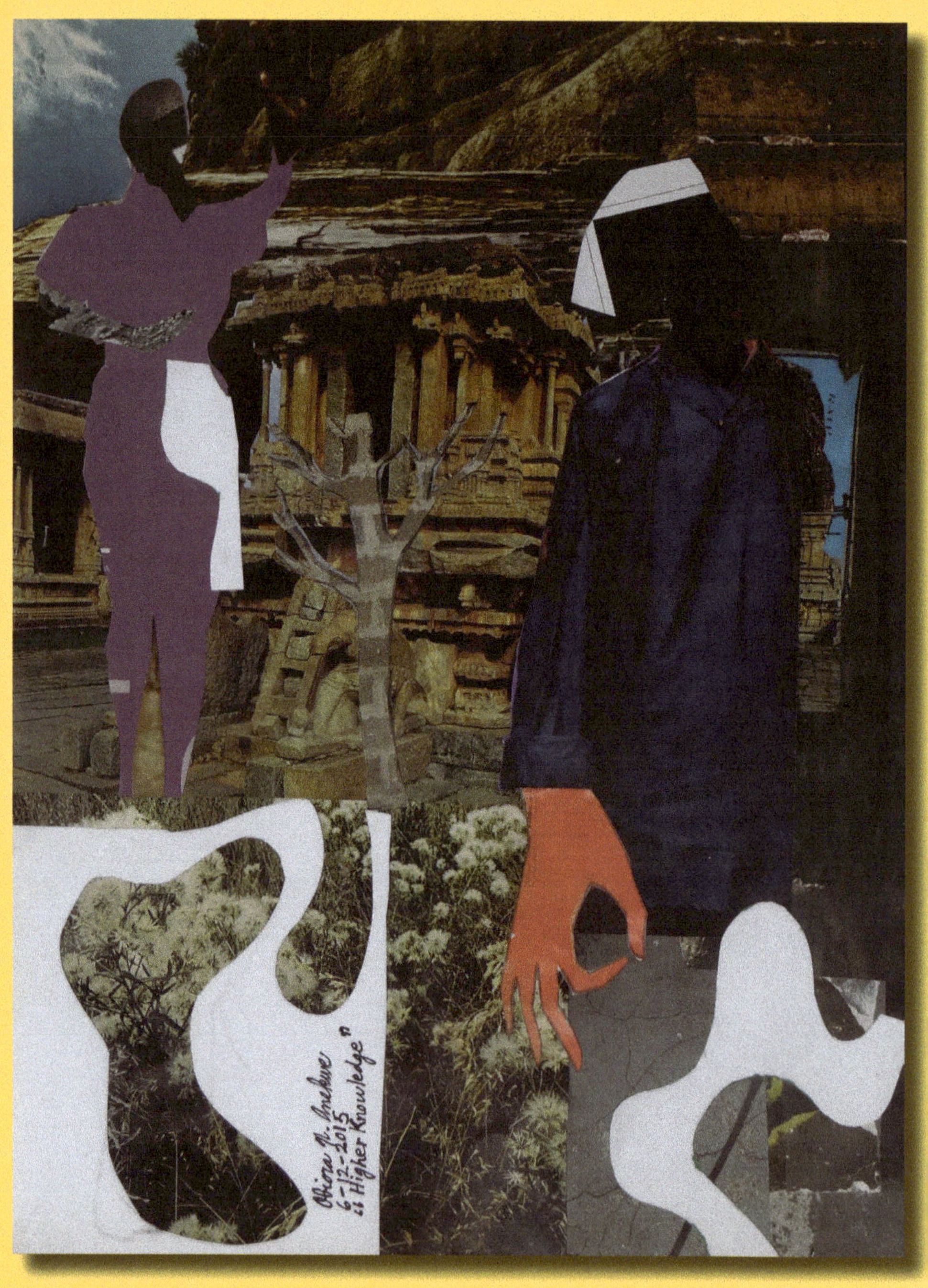
Obiora N. Anekwe
6-12-2015
"Higher Knowledge"

Higher Knowledge

One of the purposes of the Willowbrook State School was to educate students with special needs in a residential learning setting. Unfortunately, there is enough evidence to prove that educating these students was not paramount to their success at Willowbrook. To the contrary, many students were treated with disdain and ignored with little effort made to develop students' innermost talents. The collage highlights what higher knowledge should look like through abstract bodily forms. The entire image portrays ancient cultures of the world as civilizations that emphasized education as a way of life, intertwined with societal norms. The two human-like figures represent intellectualism at its height, especially the right hand figure which shows a being with long braided hair. The length of the hair represents maturity and wisdom in ancient society. The confidence of each figure is evident, especially in how they each hold themselves up in high esteem, respect, and dignity for human integrity. Although some people in our modern society reject intellectualism to a great degree, it can be recaptured by looking at past societies in order to build future rational societies for human growth and development. As such, all people, in spite of physical or mental challenges, have to be accepted in a systematic society in order for it to be morally and ethically viable. In essence, the establishment of such a society is what we can essentially describe as higher knowledge.

Obiora N. Anekwe
c. 2015
MIXED MEDIA ON PAPER
12 by 16 INCHES

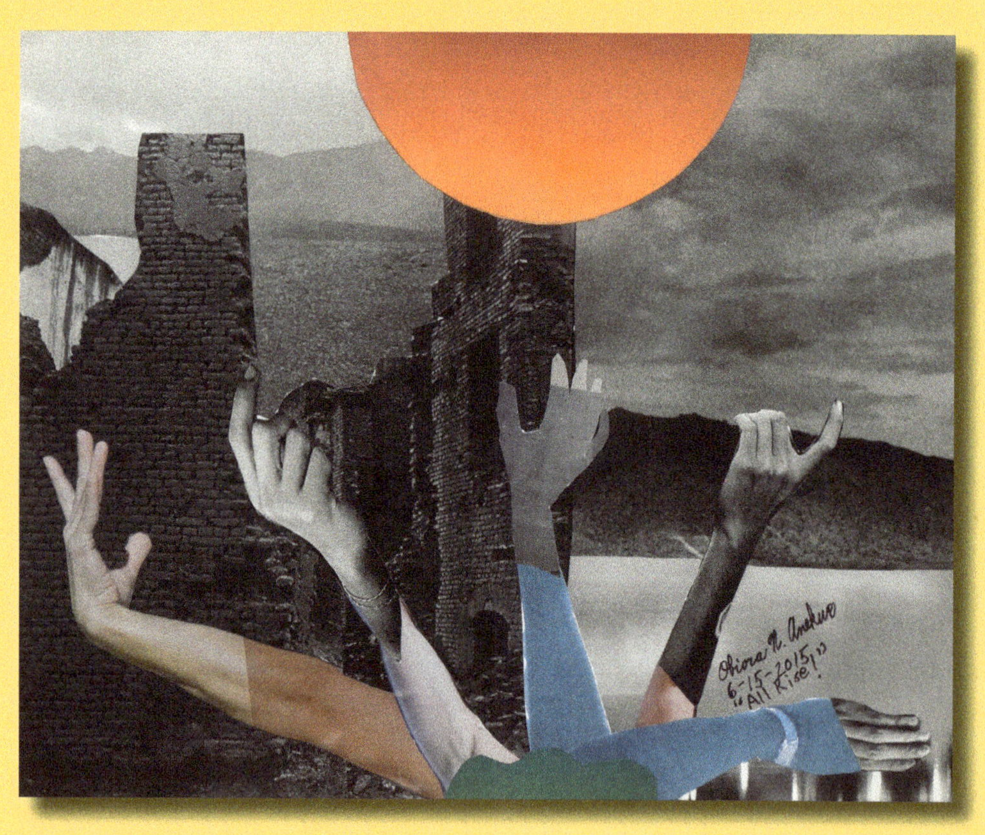
Obiora N. Anekwe
6-15-2015
"All Rise!"

All Rise!

My collage represents the symbolic rise of hope after a human atrocity. One such atrocity occurred at the Willowbrook State School. The uplifted hands in the piece represent an enduring light out of the depths of a barren earth. The background shows decaying building landscapes that have eroded out of despair. The multicolored hands represent the fact that human pain and suffering can occur to any person or group of people regardless of skin color. Ironically, the bright orange sun shines brightly in contrast to the dark skyline and melancholy mood of despair depicted in the piece.

Obiora N. Anekwe
c. 2015
MIXED MEDIA ON PAPER
$10^{1/2}$ by $13^{1/2}$ INCHES

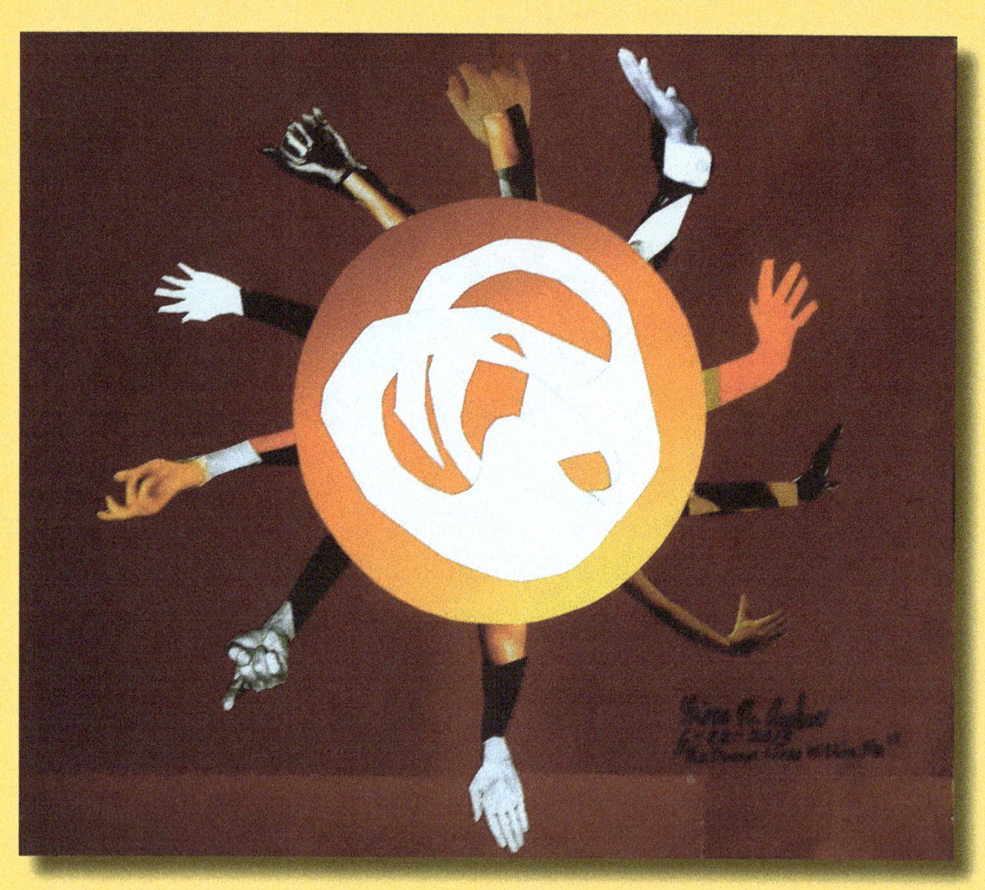

The Dream Lives Within Me

Every year during the month of January, the birthday of Martin Luther King, Jr. is celebrated throughout the nation. King fought for the civil rights of African Americans until his untimely assassination on April 4, 1968. He not only supported the Civil Rights Movement domestically, but he tirelessly fought for the human rights of all people globally.

The collage I created reflects King's vision of helping the most disenfranchised, especially those who are unable to speak for themselves, such as children with special needs. My collage also echoes the profound belief that the dream King had for our world to live in peace and justice has the possibility of existing within each person's conscious spirit. The collage's center image shows a crown slightly tilted in order to symbolize that each human being has a crown of thorns or burden that must be carried. The extension of hands surrounding this crown's circular exterior represents the various hands of humanitarians who have worked to empower the world, including the hands of Martin Luther King, Jr., Michael Jackson, Malcolm X, and Jorge Cardinal Bergoglio.

Obiora N. Anekwe
c. 2015
MIXED MEDIA ON PAPER
11 by 13 INCHES

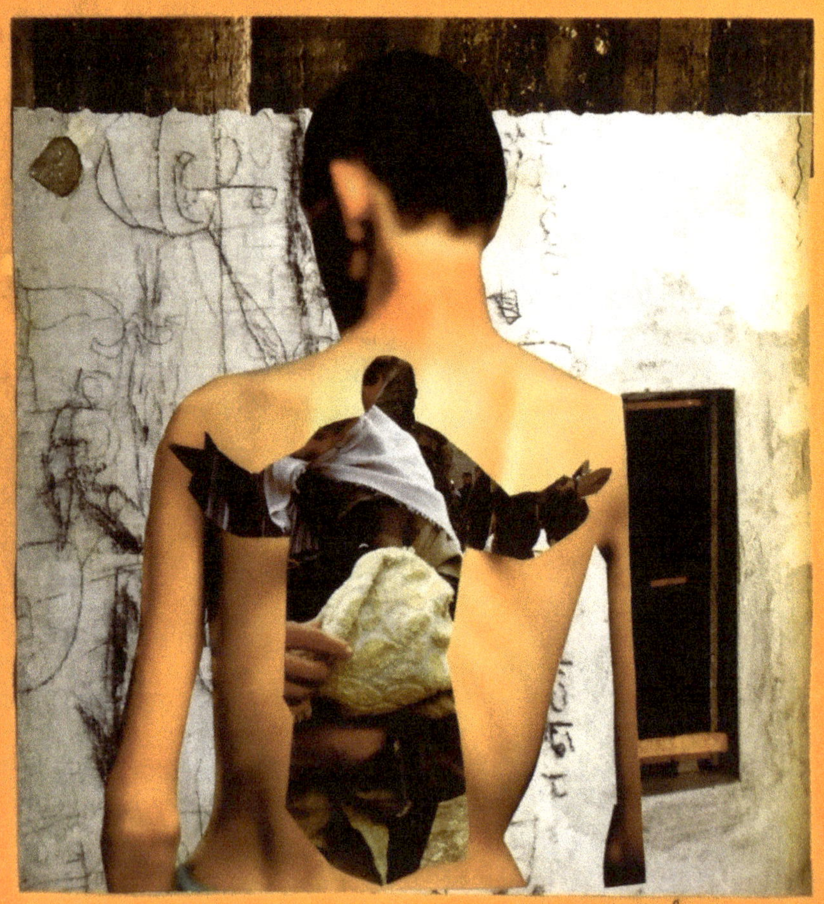

Obiora N. Anekwe
5-30-2015
"Let Us Pray"

Let Us Pray

When we pray, we often pray for areas of our lives that affect us in the most direct way. But in the piece, *Let Us Pray*, I am placing emphasis on the fact that prayer requires a collective effort for change to really occur. As such, a collective spiritual force guided with supreme consciousness reveals tangible solutions to our most challenging global issues. One global issue that has continued to hinder human progress is that of international ethnic cleansing, neo-eugenics, global genocide, and white supremacy based on artificial constructs of race and intelligence. As reflected in the artwork, tribal and religious ethnic cleansing wars have been prevalent in Eastern Europe between both Muslim and Christian populations, along with other aligned groups. The outcome for people in these tribal groups have been the same for many children who were housed in horrific conditions at Willowbrook.

The backsided image of a malnourished man facing an abandoned home shows him within deep contemplation and reflection during the Yugoslav Wars in Eastern Europe. On the back of the man is an image of a religious priest who stretches his hands for prayer. His posture is a call for prayer and subsequent action by all people who understand that, as Martin Luther King, Jr. stated, "injustice anywhere is a threat to justice everywhere."

Obiora N. Anekwe
c. 2015
MIXED MEDIA ON PAPER
11 by 15 INCHES

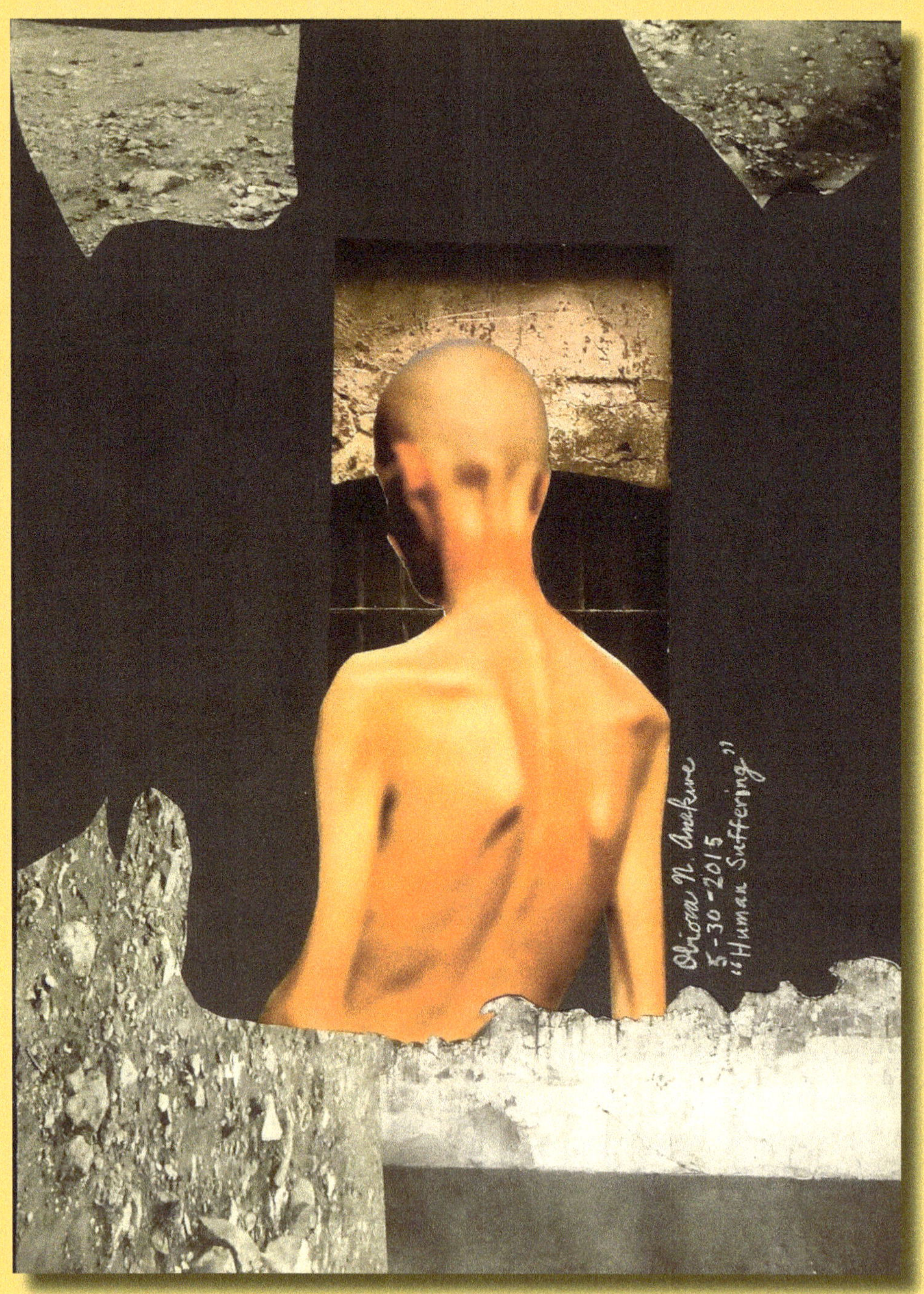
Obiora N. Anekwe
5-30-2015
"Human Suffering"

Human Suffering

Suffering is universal and transparent. It can occur to anyone during any state of one's human existence. Hence, it is significant to remember that we are all subject to the controlling substance and unique stature of human suffering. It is even more fatal and life-transforming when it damages the lives of children. In the case of Willowbrook, young men and women were used in human experiments that lacked proper informed consent from parents who entrusted doctors to essentially *do right* by their children. Instead, some state-sponsored residential mental and physically disabled facilities were used as human warehouses that medically used residents in clinical experiments.

The mixed media work entitled, *Human Suffering*, correlates how universal and transparent human suffering can actually be. This artwork also refers back to global ethnic cleansing discussed in the summary of its counterpart, *Let Us Pray*. The background shows prison bars, representative of an imprisoned soul trapped in a mournful state, paralyzed by mental and physical subjugation. The black and white stone-like structures that surround the central image of a malnourished man turning his back is symbolic of the fact that he has turned his back on the world, relegating himself to the belief that he is doomed without human interaction to bring him out of a forced state of human despair. The dreary colors used throughout the work also highlight how color sets a mood of resignation to human suffering.

Obiora N. Anekwe
c. 2015
MIXED MEDIA ON PAPER
11 by 15 INCHES

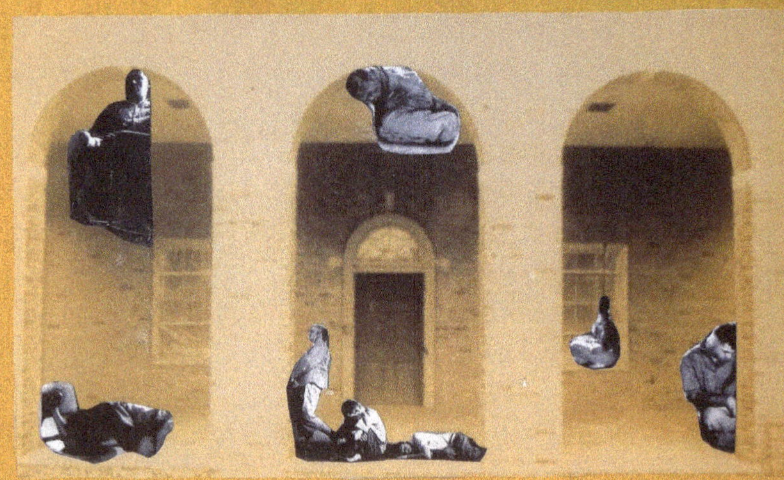

Obiora N. Anekwe
5-17-2015
"Ghosts of Willowbrook"

Ghosts of Willowbrook

The ghosts or spiritual beings of the Willowbrook State School transcendently live within and through the environment their physical bodies once occupied. In my collage, I want to emphasize this very point by showing black and white images of oppressed and depressed physical bodies of people who once resided at Willowbrook decades ago. Those images emphasize the fact that although the medical studies that occurred at Willowbrook were historically of the past, such unethical experiments still have a validity toward showing that the past surely matters and can be repeated if lessons are left unlearned. Interestingly, the black and white cut-out human images physically project out from the dark and dreary windows of the building. This symbolizes that human life must take precedence over material gain and profit. The earth tone colors found within the collage reflect the environmental depiction as shown in the physical landscape and atmosphere of the former Willowbrook State School.

Obiora N. Anekwe
c. 2015
COLLAGE ON PAPER
12 by16 INCHES

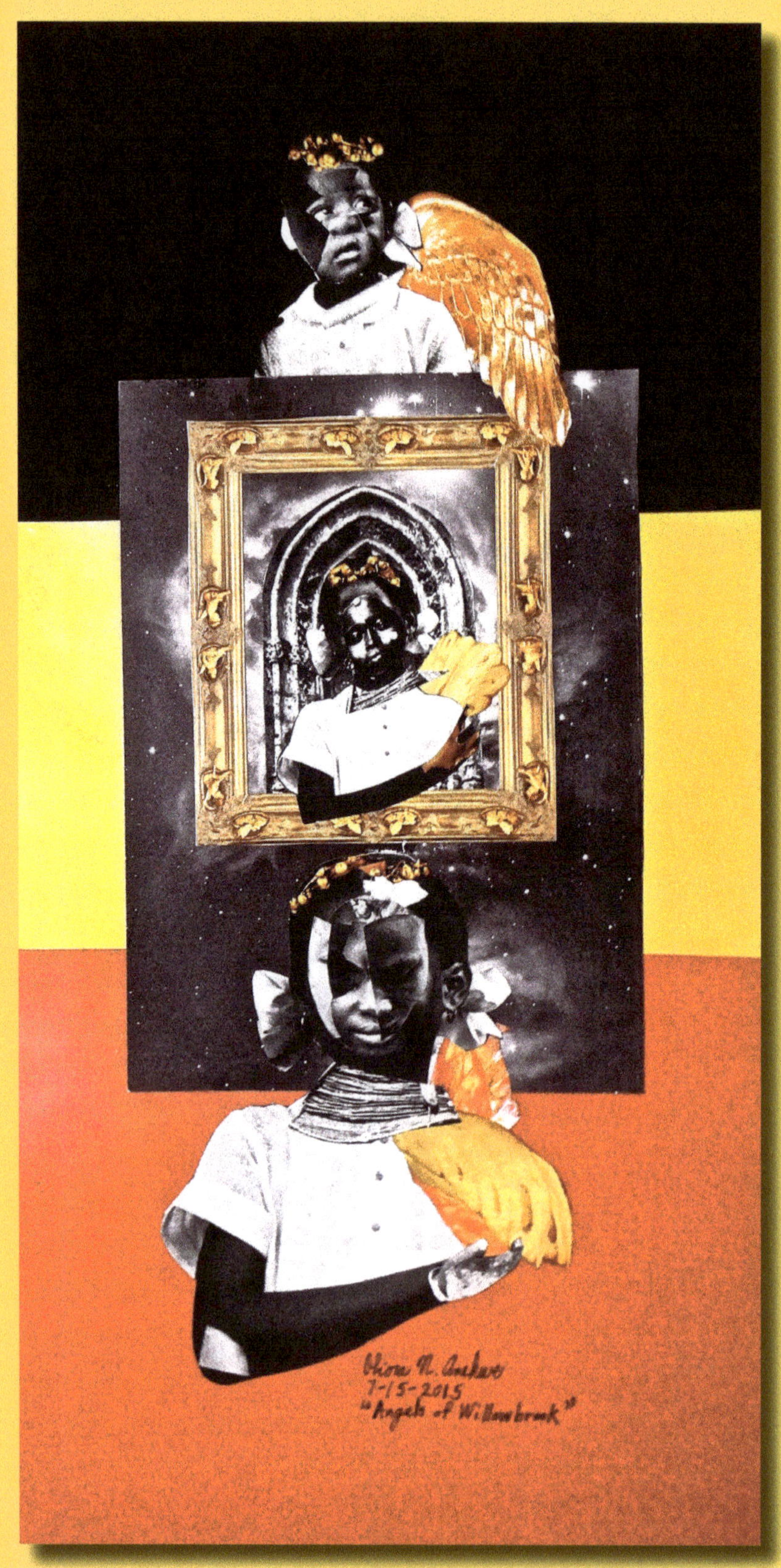
7-15-2015
"Angels of Willowbrook"

Angels of Willowbrook

Within many ancient cultures, angels were considered messengers from God or the Collective Spirit. In modern times, angels still serve the same purpose. *Angels of Willowbrook* is an example of how impactful messages of healing and reconciliation can be translated through angelic beings like the three girls portrayed in this collage. These angels' message is simple: we are all children of God who deserve respect and human dignity. As such, our own human shortcomings are a means by which the Most High can be glorified.

The background colors of red, black and yellow correlate with colors found on the German flag, signaling awareness of the German Holocaust of Jewish children during World War II. These colors also symbolize the unification of Germany after the Cold War.

At Willowbrook, many residents were only seen as research contributors to scientific advancement for experiments held there. My collage image reminds its viewer that these children at Willowbrook were more than contributors to science. Rather, they were human beings with the same inalienable rights as other people who live and occupy space in our national and global society.

Obiora N. Anekwe
c. 2015
MIXED MEDIA ON PAPER
$12^{1/2}$ by 25 INCHES

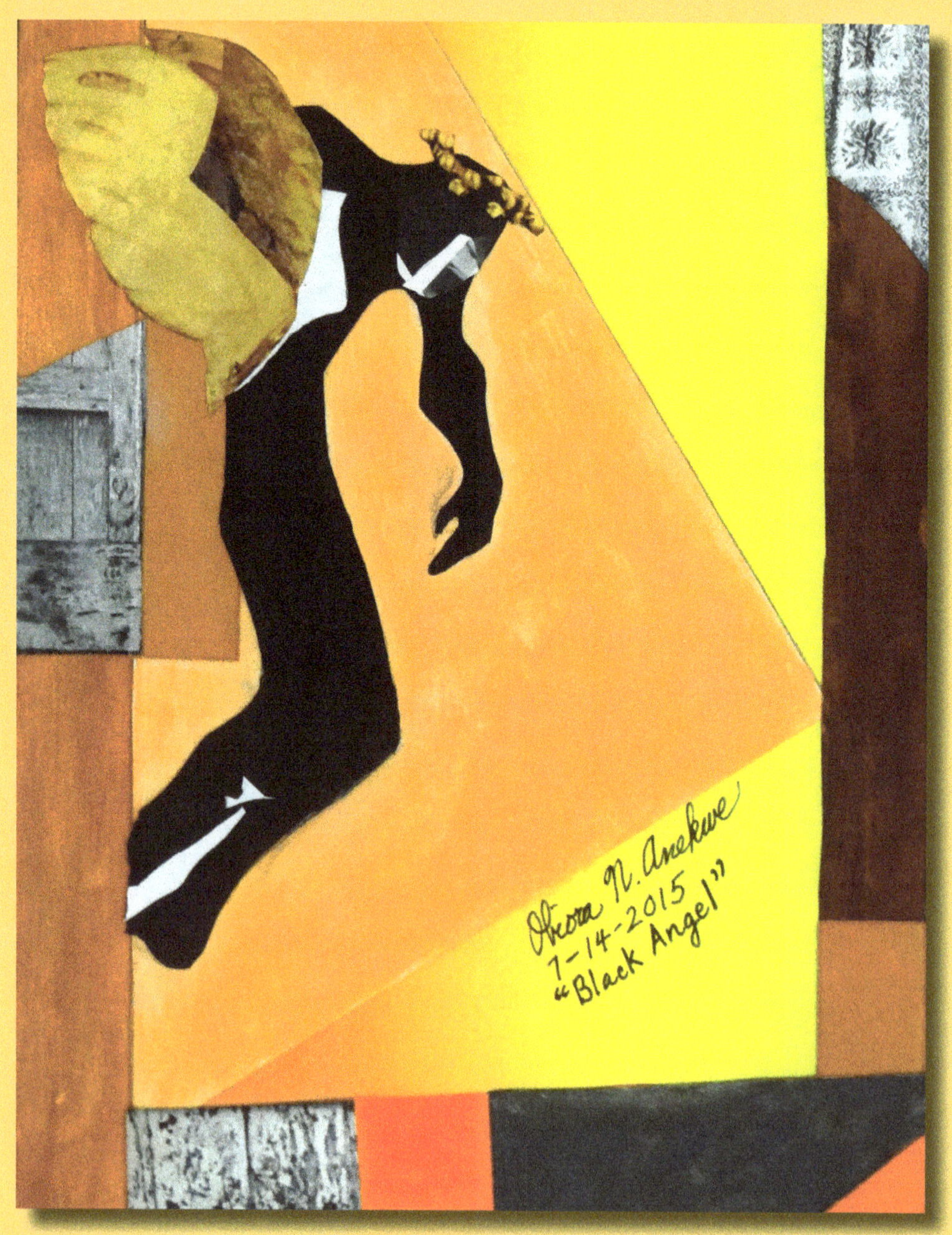
7-14-2015
“Black Angel”

Black Angel

Angels are heavenly beings who are messengers that move and conduct actions out of loyalty and duty to the Most High. They are also warriors charged with protecting God and his children from harm and evil. In many cases, human beings have been protected by angels even when negative circumstances seem to deter and hinder God's children. In the case of Willowbrook, a spiritual battle existed in which human beings charged with helping our most vulnerable children were essentially doers of evil who abused and misused their call to serve humankind. Nevertheless, God's angels still watched over the children of Willowbrook because some survivors of medical experiments at the institution were able to recall, voice, and expose what happened to them as a result of inhumane negligence.

In my collage, *Black Angel*, I use memory as a source of strength. The angelic being depicted in the collage takes on a philosophical and physical stature as a romanesque figure who ponders and internalizes the burdens of psychological shame residents of Willowbrook may have felt. He is a black angel whose color serves as a neutral contemplation, allowing judgment for past sins to be determined by the Most High. As the angel reflects, he holds to the belief that God is the only One who can judge wrongdoers and he, the angel, is the one charged by the Holy Spirit to translate a message of hope to God's people in spite of circumstance.

Obiora N. Anekwe
c. 2015
COLLAGE ON PAPER
8 by 11 INCHES

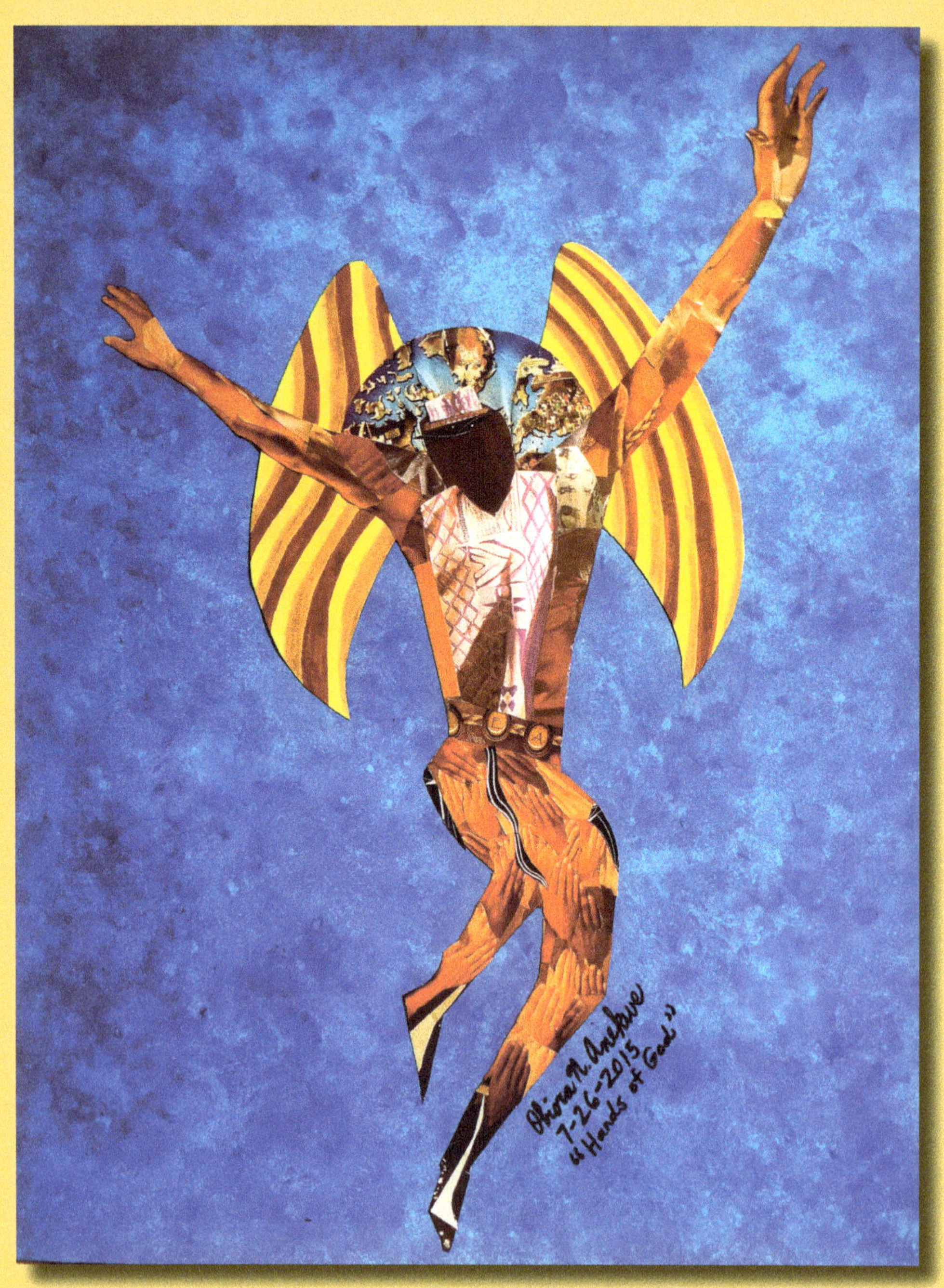
7-26-2015
"Hands of God"

Hands of God

The collage entitled, *Hands of God*, evolved organically during the summer of 2015 as a form of reflective healing and reconciliation, both personally and spiritually. The image not only symbolized my own interpretation of Willowbrook's past, but it also served as a means to use art as a therapeutic dimension to create global healing from universal suffering. The collage's title emphasizes the fact that we all serve as the hands of God, capable of making change a reality through collective efforts and discovery of self. We, as human beings, are charged with the universal duty of stopping injustice when our own conscience leads us to correct a wrongdoing perpetrated against other people who cannot speak for themselves, as was the case at the Willowbrook State School.

The angelic image's physical body in the collage is made mostly from hands from around the world, especially those of the darker race of people. Most notably, the legs of the angelic being are encompassed with upward hands to symbolize the collective effort of lifting up justice and righteousness. The angelic form also represents Jesus the Christ. He is depicted as a beautifully rendered black angel with Afrocentric nuances, most noticeably in the Kente cloth-inspired crown Christ wears in the image. Above the crown, the black male body of Christ shoulders the planet Earth. Within the global image of Earth, small faces of children are faintly visible to the naked eye. These children remind us that all God's children, including those who are victimized by unethical clinical trials, are lifted up in God's eyes, both physically and spiritually.

Obiora N. Anekwe
c. 2015
MIXED MEDIA ON PAPER
19 by 24 INCHES

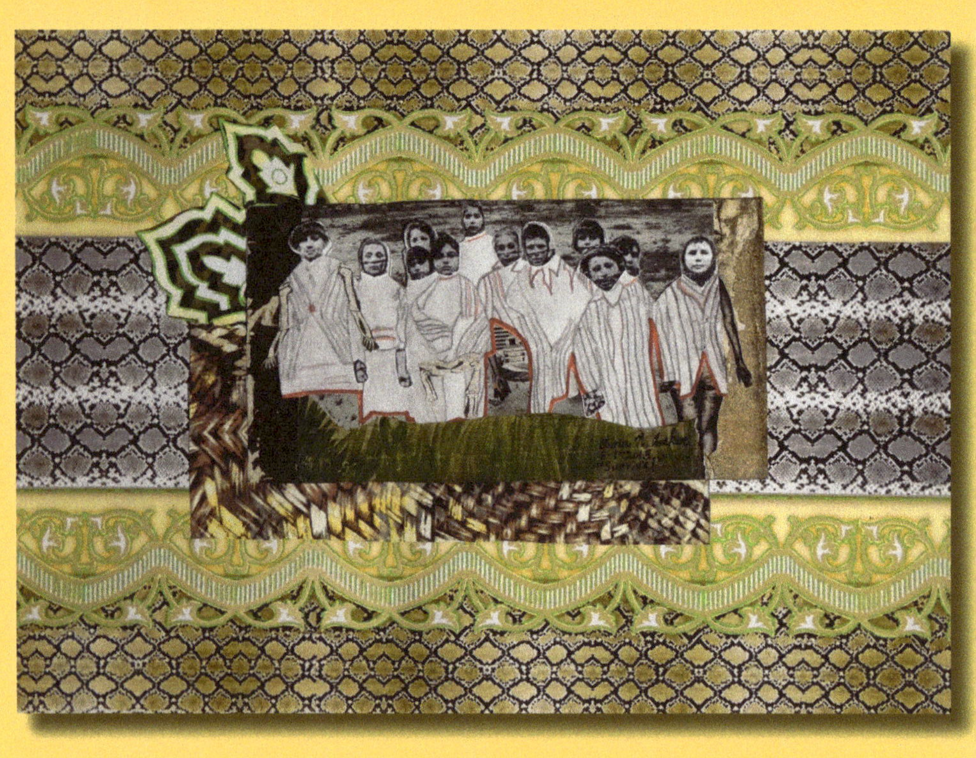

Survival

Global human atrocities have parallel ethical ramifications that affect children with both unintended and intended consequences. As a result of this, many artworks have documented universal themes of oppression with a need for global justice. The collage entitled, *Survival*, is no exception. As the viewer looks at the faces of Jewish children in the collage, one can observe that innocence is prevalent in their eyes. But more tragically, one can visibly see the skeletal figures these children possess as a result of malnutrition and starvation imposed by the Nazi-ruled government of Germany during World War II. The image correlates with what can be imaged as survival at its best for children who also lived at the Willowbrook State School, a type of human concentration camp in and of itself as well.

On the other hand, the natural color scheme and usage of greens, browns, yellows, reds, and grays in the collage hints to the fact that natural environmental locations such as Willowbrook and the Auschwitz-Birkenau concentration camp can hide the reality of abuse perpetrated upon vulnerable children. It would never have been suspected that human experiments were held on our most vulnerable in such tranquil and peaceful external environments as Willowbrook and the Auschwitz-Birkenau concentration camp. But behind the layers of such environmental tranquility lurks medical evil. The burdensome eyes of our youth hide the darkness of human suffering.

Obiora N. Anekwe
c. 2015
COLLAGE ON PAPER
$11^{1/2}$ by 16 INCHES

THE SPECIAL OLYMPIC GAMES

There are six art collages in this section, focusing on the U.S. Special Olympic Games. These collages connect to how the Special Olympics became an avenue by which many formerly abused special needs students could be encouraged and inspired to achieve their goal of physical excellence through sports. These art collages are self-explanatory and make a positive impact in how we view children with special needs. The value of their own physical fitness and well-being is just as pertinent to their self-esteem and self-concept as any other child who strives to achieve physical excellence.

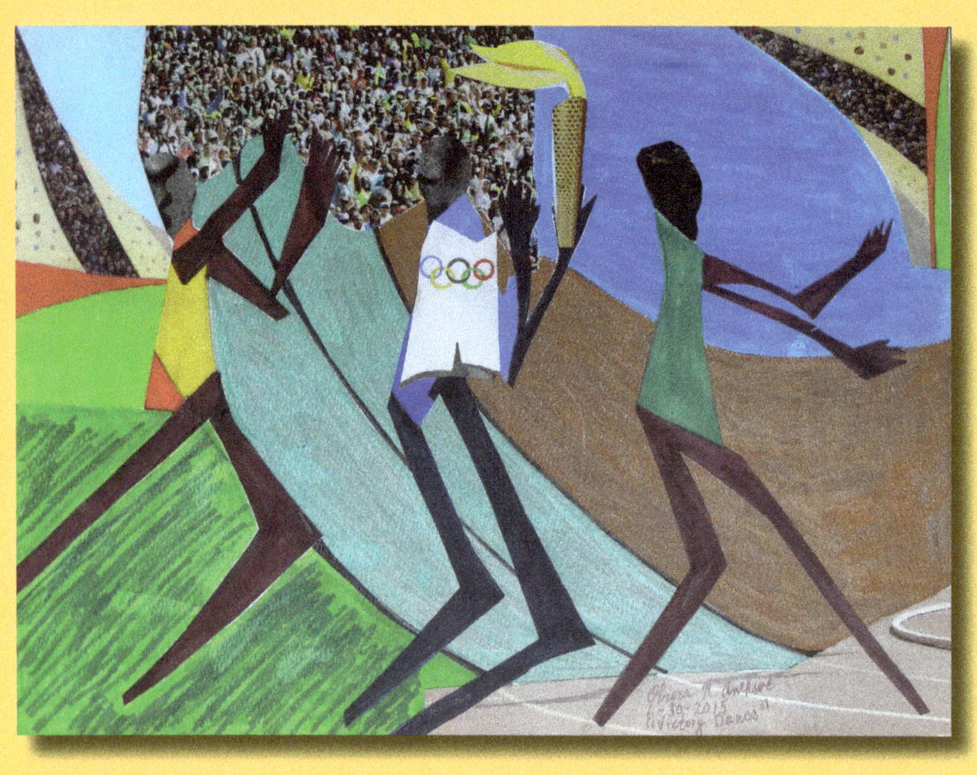
"Victory Dance"

Victory Dance

When people win big in life, there is a human tendency to celebrate. In my collage entitled, *Victory Dance*, I wanted to show how children with special needs react when they achieve a sense of victory, whether it be physical, spiritual, or psychological. Often, the accomplishments of children are overlooked and not acknowledged. Visual images such as this one often remind us that each moment of accomplishment deserves acknowledgement. Globally speaking, dance is a representation of different types of celebrations, whether it be birth, marriage, or death. Dance especially expresses a people's inner joy and deepest emotion. It is also one of the most therapeutic attributes of expression that releases tension, while stimulating the brain for one's ultimate creativity. In many ways, dance is arguably the highest physical art form in which human emotion is connected to visual and bodily expression. As for this collage, the vivid nature of colors used in my piece is a reflection of the inner feelings of victory children feel when they compete in the Special Olympics.

Obiora N. Anekwe
c. 2015
COLLAGE ON PAPER
11 by 15 INCHES

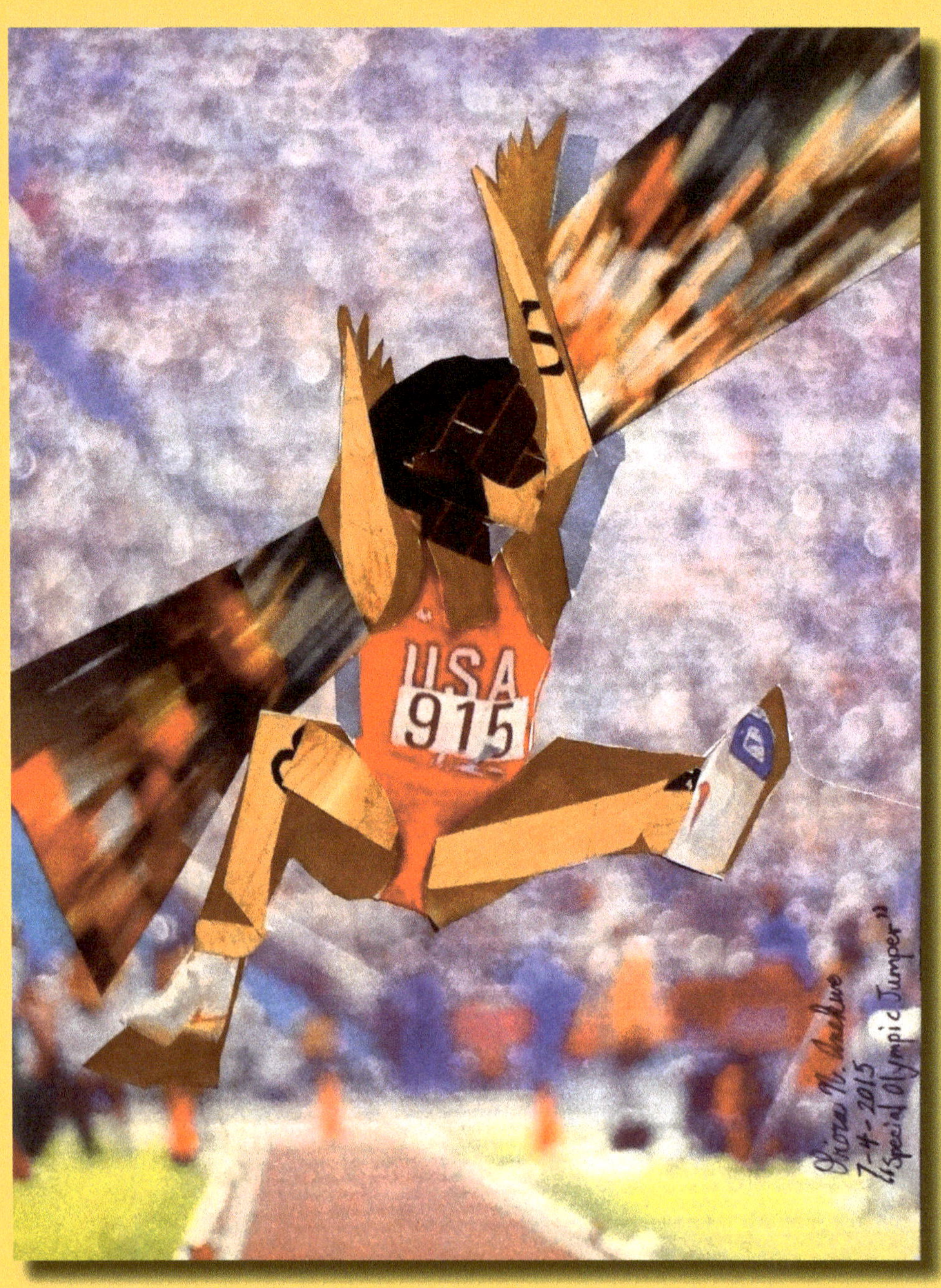
USA
915
7-4-2015
"Special Olympic Jumper"

Special Olympic Jumper

The collage image of the Special Olympic jumper is layered with multiple colors on his arms and legs, speaking to the ethnic and coloristic diversity of children with special needs. The collage's background shows a blurred image of the cheering crowd who are undoubtedly supporters of the Special Olympics. The wings on the athlete are vivid and spectacular in physical range and scale, symbolizing the angelic value and nature of children who have special needs. Overall, I created this collage to emphasize that all children have to be provided the opportunity to reach their goals, no matter what barriers society puts in place to limit their progress.

Obiora N. Anekwe
c. 2015
COLLAGE ON PAPER
11 by 15 INCHES

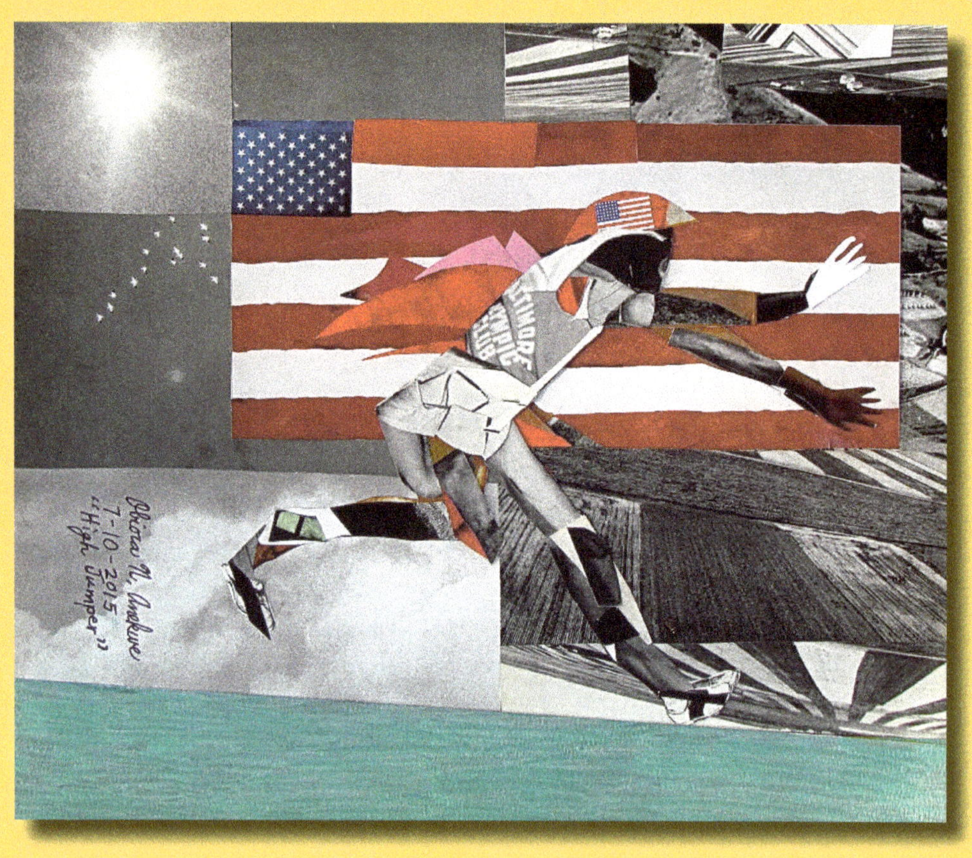
Obiora N. Anekwe
7-10-2015
"High Jumper"

High Jumper

In life, it is a natural desire to want to rise to the highest level of achievement and accomplishment. Everyone wants to be in a better place in order to become a better person. In the collage entitled, *High Jumper*, a higher level of success is portrayed through the medium of sports. The symbol of a sports jumper shows that all people have the ability, no matter the perceived limitation or challenge, to become winners in life. We are even capable of overcoming or jumping over hurdles and challenges that may seem unwinnable.

My collage is a mixture of paper cutouts from magazines and books that serves as a symbolic admixture of colors and shapes from unconscious and conscious thoughts about my physical, spiritual, and psychological perceptions of perseverance and persistence. In other words, life's deepest challenges can make us better human beings who are capable of becoming more compassionate toward those in dire circumstances. As a result, the most vulnerable are not just looked down upon, but they are instead viewed through the eyes of compassion for one's fellow citizen.

Obiora N. Anekwe
c. 2015
COLLAGE ON PAPER
12 by $14^{1/2}$ INCHES

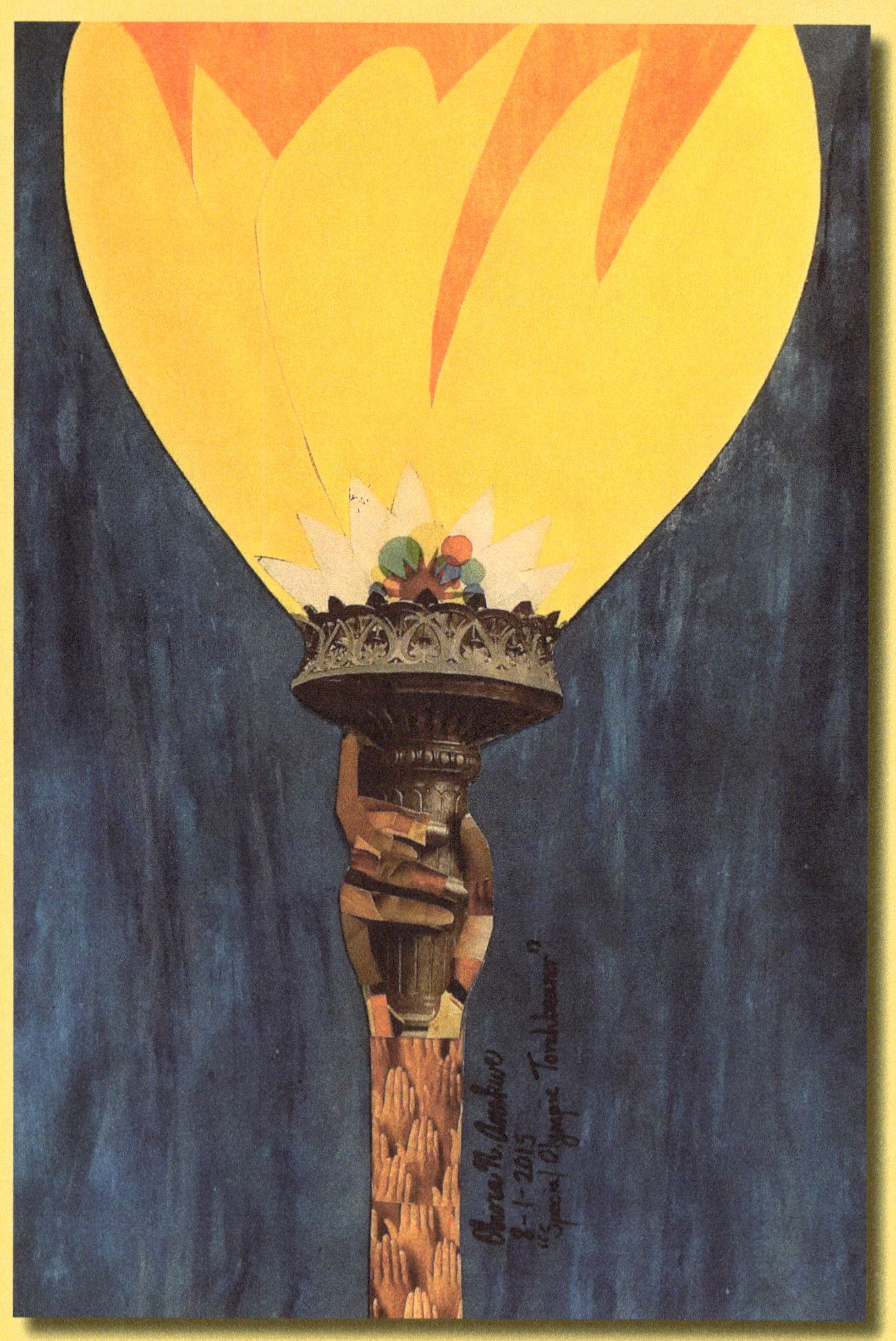
8-1-2015
"Special Olympic Torchbearer"

Special Olympic Torchbearer

The Special Olympics is the world's largest sports organization for children and adults with intellectual and physical challenges. The games are held year-round and provide training and facilitate competitions for more than 4.5 million athletes in 170 countries. Special Olympics competitions are held every day, all around the world-including local, national, and regional competitions.

The history of the Special Olympics began when, in June 1962, Eunice Kennedy Shriver started a day camp for children with intellectual and physical challenges at her home in Potomac, Maryland. She started her camp because she was deeply concerned about children with intellectual and physical challenges who had no physical place to play competitive sports. Using Camp Shriver as an example, Eunice Kennedy Shriver, who was head of the Joseph P. Kennedy, Jr. Foundation and part of President John F. Kennedy's Panel on Mental Retardation, promoted the concept of involvement in physical activity as a means of social inclusion. As a result, Camp Shriver soon became an annual event as the Kennedy Foundation gave grants to universities, recreation departments, and community centers to hold similar camps.

Although the Special Olympics has had numerous logos and symbols, it was important for me to create one that emphasized the diversity of races, genders, and creeds of people who participate in the games. The collage entitled, *Special Olympic Torchbearer*, serves as a symbolic tool to represent the true diversity of children and adults who are affected by mental and physical challenges. The collage of hands on the front of the torch shows that all people throughout the world can come together peacefully in order to play competitive sports in a challenging and productive manner. The Special Olympics serves as a model as to how universal sports participation is both realistic and tangible globally. The bright colors from the top flame of the torch are symbolic of a renewed vigilance and hope that humankind can unite for the common cause of supporting those who have intellectual and physical challenges.

Obiora N. Anekwe
c. 2015

COLLAGE ON PAPER
19 by 24 INCHES

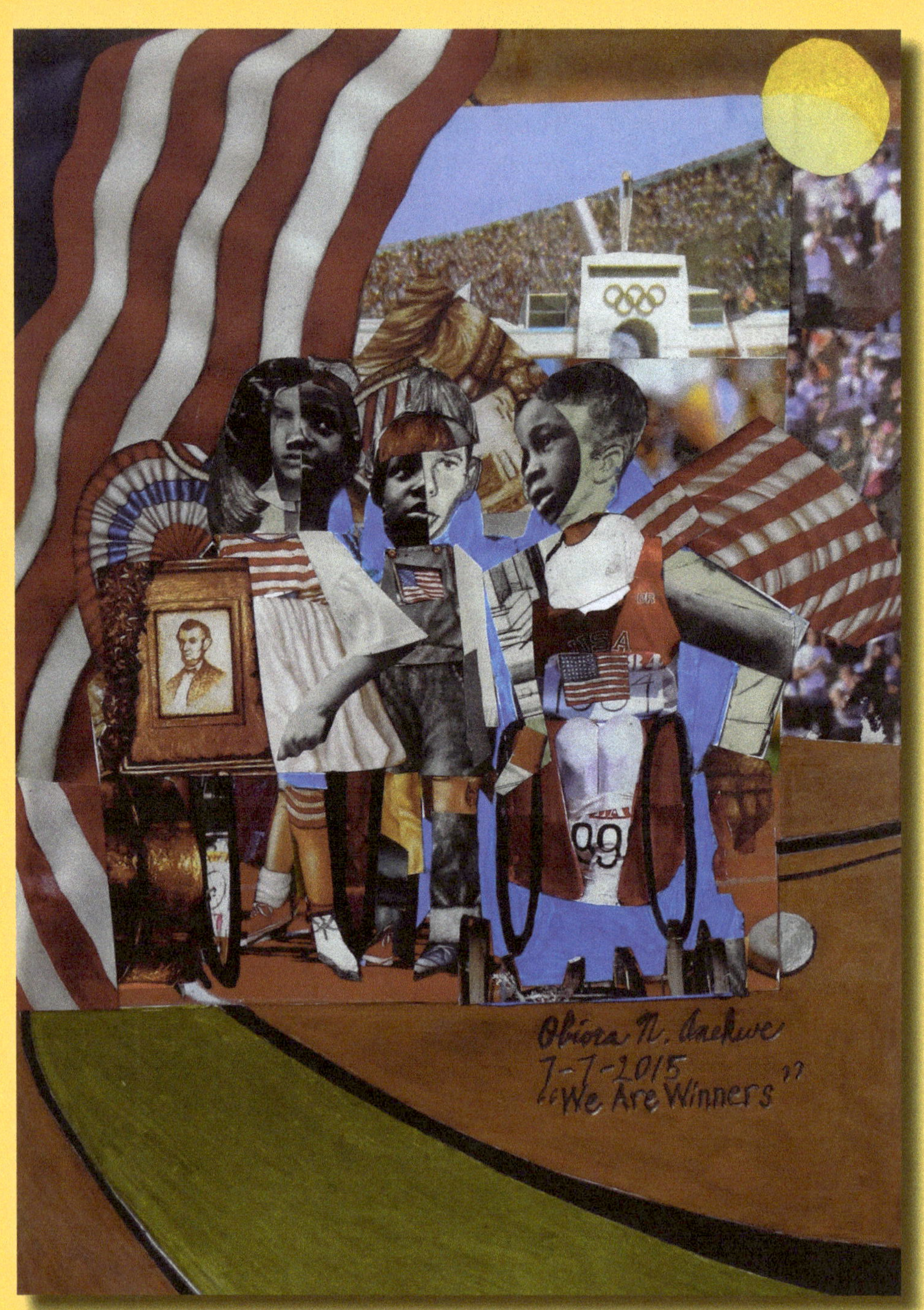
USA
99
Obiora N. Anekwe
7-7-2015
"We Are Winners"

We Are Winners

My collage entitled, *We Are Winners*, encompasses several layers of meaning. For example, the Olympic archway in the upper right hand corner reflects the explicit ideal that we are all winners. It also symbolizes the entry point to human potential and possibilities absent of discriminatory practices and injustice. The collage also reminds us all that it is perseverance and persistence that determines the winners of life's obstacles. The three children in the collage's center accentuate the diversity of children who participate in the Special Olympics. These children are compositions and layers of multiple images gathered from magazines and other discarded documents. The young lady holding the picture of Abraham Lincoln on the left side reminds us that even our greatest leaders have had to overcome immense adversities to become successful. In particular, President Lincoln had to overcome several childhood illnesses and clinical depression as an adult. As art viewers reflect on the overall composition of this collage, they will hopefully be reminded that to overcome adversity is the beginning of life's victory!

Obiora N. Anekwe
c. 2015
COLLAGE ON PAPER
$8^{1/2}$ by 12 INCHES

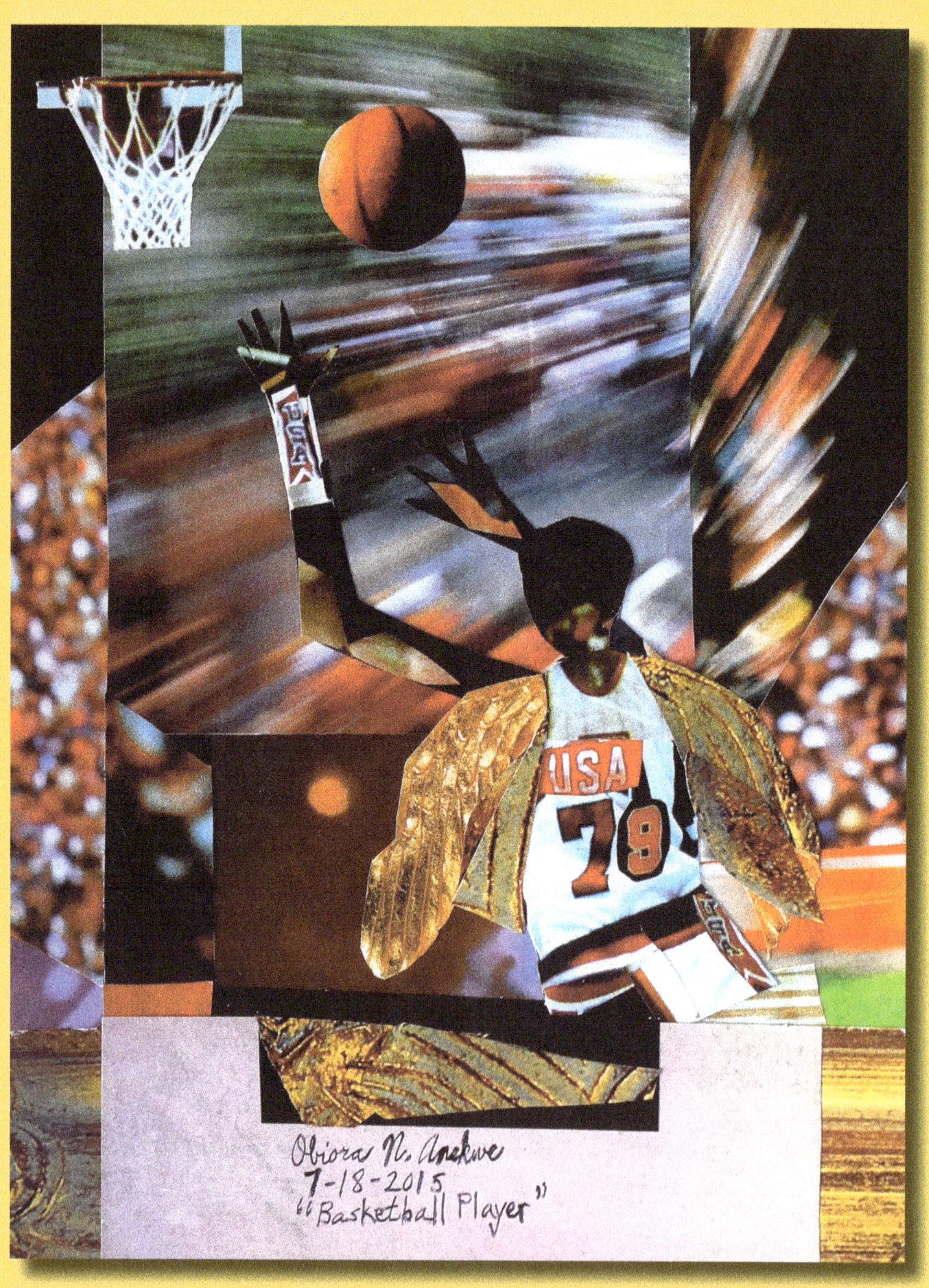
USA
USA
79
Obiora N. Anekwe
7-18-2015
"Basketball Player"

Basketball Player

Like many games in North American history, basketball was once a segregated sport. It was not until basketball began to integrate that African Americans dominated the sport. Today, young people with disabilities are also given the opportunity to participate in playing basketball through the Special Olympics. They are now extended the same sports' rights as other minority groups who were also denied the right to play basketball. The collage image entitled, *Basketball Player*, exemplifies how sports can serve as a positive and transformative form of recreation and growth. It serves as a confidence builder for young people who desire to believe in something bigger than themselves. But more strikingly, baseball is a metaphor for life that reminds the viewer that achievement is possible point by point in the game of life. Although accomplishing one's goal takes time, the game of basketball can serve as a great motivator that demonstrates how hard work and perseverance pays off. This type of symbolism is not only significant to children with mental or physical challenges, but it is also imperative for all of us to remember and cherish.

Obiora N. Anekwe
c. 2015
COLLAGE ON PAPER
13 by 10 INCHES

EDUCATIONAL AND SOCIAL JUSTICE

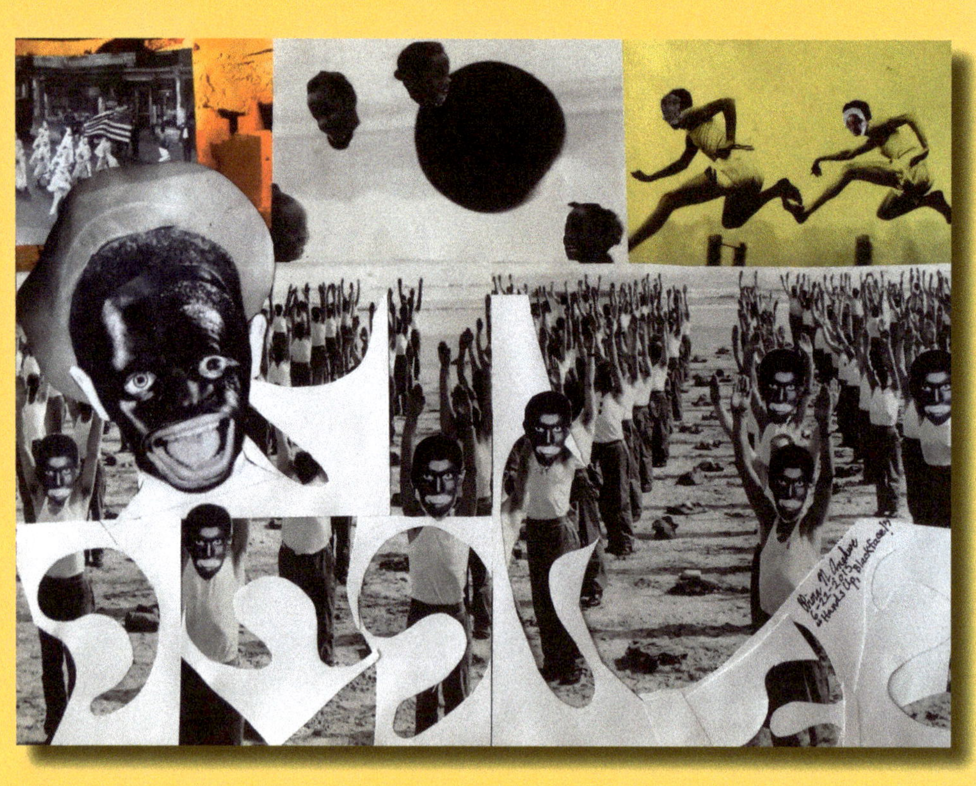
6-22-2015
"Hands Up, Blackface!"

Hands Up, Blackface!

Hands Up, Blackface! correlates with the fact that the Special Olympics came into fruition as a backlash against unethical human experiments held on campus at the Willowbrook State School. The collage shows African American Olympic runners in the far right while Ku Klux Klan men rally in opposition to civil rights for African Americans. The other images focus on the four black girls killed in a church bombing in Birmingham, Alabama during the Civil Rights Movement. Coincidently, the larger portion of the collage image at the bottom shows men in blackface exercising in an expansive landscaped area. This specific image relates back to how African American men are now being unjustly incarcerated in the American prison system. Overall, the collage highlights the fact that some of the same human injustices that were perpetrated on residents at Willowbrook are now being conducted on minorities who occupy the American prison system, such as the purposeful recruitment of prisoners in unethical clinical trials.

Obiora N. Anekwe
c. 2015
COLLAGE ON PAPER
$18^{1/2}$ by 14 INCHES

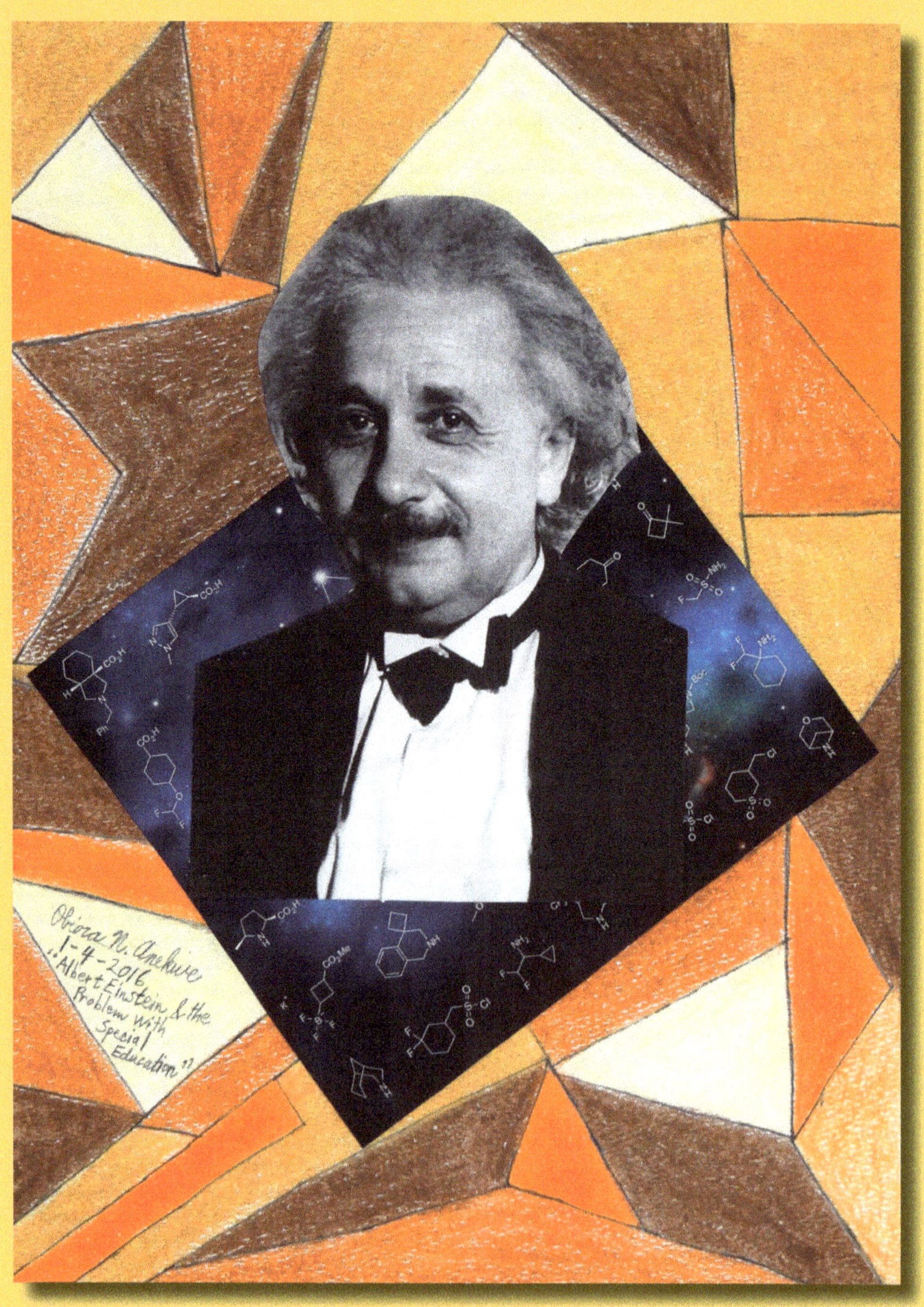
Obiora N. Anekwe
1-4-2016
"Albert Einstein & the
Problem with
Special
Education"

Albert Einstein & the Problem with Special Education

Albert Einstein (1879-1955) was known as a true genius and one of the world's greatest thinkers in physics and mathematics. He was also a great humanitarian, researcher, and teacher who taught aspiring physicists and mathematicians at Princeton University until his death. As depicted in my collage, Einstein looked beyond himself to contribute to the world. His work was important to him, but many scholars considered his pursuit of justice for all people as the greater role. The background collage image of scientific renderings shows the complexity of scientific exploration in the universe. The flat geometric images I created were hand drawn without any assistance of a ruler. The colors are earth tone in nature to symbolize the down-to-earth nature of Einstein's personality. These tones were created from oil pastels. The reason I wanted to include this particular collage in my book is because it highlights the fact that although Einstein was considered a genius, many children who had similar eccentric talents during his day were misdiagnosed as "mentally deficit" and sent to residences such as Willowbrook State School. In particular, some historians contended that Einstein was not a good student as a youth. But through research, I discovered that this was simply a myth because in Germany, the grade scaling system reversed. Einstein's grades that were interpreted as academically proficient were later incorrectly interpreted by historians as low academic grades. Unfortunately, this academic adjustment and change in Germany's pupil grading system aided in perpetuating the myth that Einstein was not a good student as a youth. It is important to note that such myths can ultimately promote a prejudicial opinion toward children who are gifted, such as Albert Einstein. As a result, I want this collage to serve as a reminder that every child has a unique gift to give to the world, even when it is not as evident as it should be to those who are charged with measuring the holistic abilities of children.

Obiora N. Anekwe
c. 2015

COLLAGE ON PAPER
11 by 15 INCHES

Photograph of Albert Einstein retrieved from C. Fisher/AMNH Library Special Collections

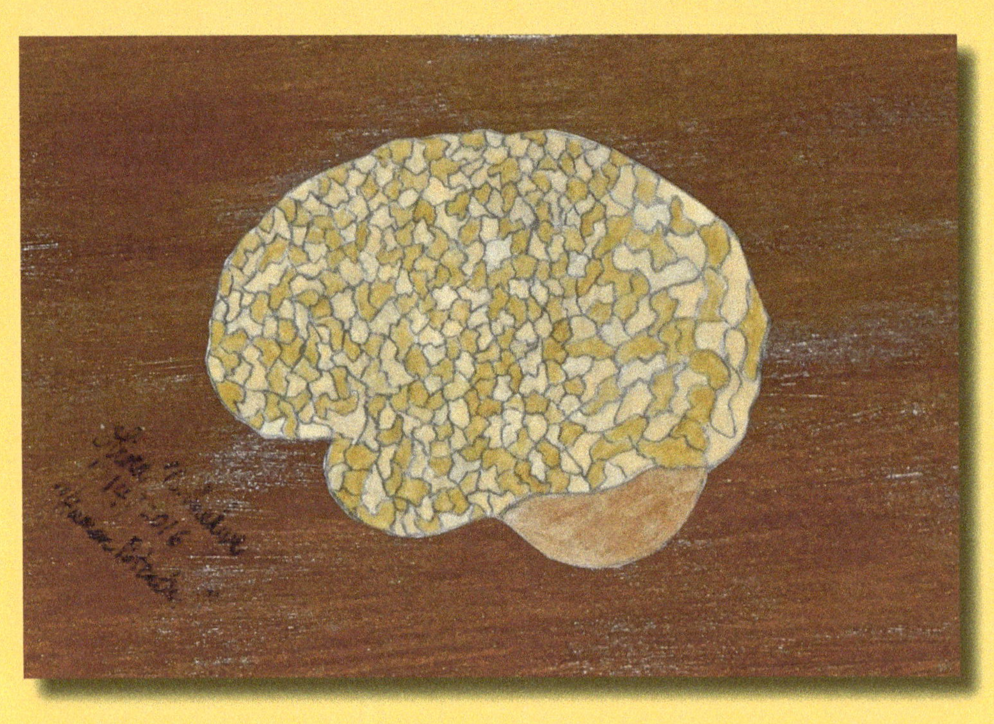

Human Potential

My collage, *Human Potential*, centers on the power of the human brain and its overall potential. Most researchers agree that the idea that humans only use about 10 percent of the brain is actually untrue. This myth has promoted an incorrect idea that most humans do not live up to their human potential when, in fact, most humans use 100 percent of their brain in everyday functioning in order to conduct complex to more mundane human activities. The image of the brain that was drawn shows the complexity and multiple parts of the brain through illustrative form. The use of naturalistic colors in illustrating the brain helps to demonstrate how one should view the power of the brain from a grounded and earthly perspective.

As so vividly demonstrated in the treatment of residents at Willowbrook, many children were disregarded and abused because society did not see their mental capacity and potential. It is unfortunate that so-called "experts" in the field of education and psychology, irrespective of their intent, tend to make irrational judgments about the mentally and physically challenged based on unrealistic social norms that are often culturally biased. The fact of the matter is that many people who were historically considered outcasts based on the lack of supposed mental functioning were, in fact, geniuses far ahead of their time in creativity and innovation. We, as imperfect human beings ourselves, have to be careful in how we label and view others based on the human functionality of the brain, because knowledge and the application of it is relative according to the time and place it is measured.

Obiora N. Anekwe
c. 2015
MIXED MEDIA ON PAPER
9 by 12 INCHES

www.ingramcontent.com/pod-product-compliance
Lightning Source LLC
La Vergne TN
LVHW051938100826
845154LV00015B/162/J

* 9 7 8 0 5 7 8 1 8 1 6 7 7 *